Genetics

FOR

DUMMIES®

2ND EDITION

by Tara Rodden Robinson

WILEY

John Wiley & Sons, Inc.

Genetics For Dummies®, 2nd Edition

Published by
John Wiley & Sons, Inc.
111 River St.
Hoboken, NJ 07030-5774
www.wiley.com

Copyright © 2010 by John Wiley & Sons, Inc., Hoboken, New Jersey

Published simultaneously in Canada

Library of Congress Control Number: 2010924590

ISBN: 978-0-470-55174-5 (pbk); ISBN 978-0-470-63446-2 (ebk); ISBN 978-0-470-63447-9 (ebk); ISBN 978-0-470-63448-6 (ebk)

Manufactured in the United States of America

15 14 13 12

WILEY

About the Author

Tara Rodden Robinson, RN, BSN, PhD, is a native of Monroe, Louisiana, where she graduated from Ouachita Parish High School. She earned her degree in nursing at the University of Southern Mississippi and worked as a registered nurse for nearly six years (mostly in surgery) before running away from home to study birds in the Costa Rican rain forest. From the rain forests, Tara traveled to the cornfields of the Midwest to earn her PhD in biology at the University of Illinois, Urbana-Champaign. She conducted her dissertation work in the Republic of Panama, where she examined the social lives of song wrens. She received her postdoctoral training in genetics with Dr. Colin Hughes (then at the University of Miami) and through a postdoctoral fellowship at Auburn University. Dr. Robinson received a teaching award for her genetics course at Auburn and was twice included in *Who's Who Among America's Teachers* (2002 and 2005).

Currently, Tara teaches genetics via distance education on behalf of the biology program at Oregon State University. On the research side, Dr. Robinson has conducted research on birds in locations all over the map, including Oregon, Michigan, Yap (part of the Federated States of Micronesia), and the Republic of Panama. Examples of her work include using paternity analysis to uncover the mysteries of birds' social lives, examining the population genetics of endangered salmon, and using DNA to find out which species of salmon that seagoing birds like to eat.

When not traveling to exotic places with her husband, ornithologist W. Douglas Robinson, Tara enjoys hiking the Coast Range of Oregon with her two dogs in training for the Susan G. Komen 3-Day for the Cure. You can find out more about her at www.thegeneticsprofessor.com.

Dedication

For Douglas: You are my Vitamin D.

Author's Acknowledgments

I extend thanks to my wonderful editors at Wiley: Elizabeth Rea, Chad Sievers, Todd Lothery, Stacy Kennedy, Lisa J. Cushman, and Mike Baker (first edition). Many other people at Wiley worked hard to make both editions of this book a reality; special thanks go to Melisa Duffy, Lindsay MacGregor, Abbie Enneking, Grace Davis, and David Hobson.

Many colleagues and friends provided help. I enjoyed lively discussions with and gained much insight into the nature of the epigenome from Jonathan Weitzman. I thank Doug P. Lyle, MD, Walter D. Smith, Benoit Leclair, Maddy Delone, and Jen Dolan of the Innocence Project; and Jorge Berreno (Applied Biosystems, Inc.), Paul Farber (Oregon State University), Iris Sandler (University of Washington), Robert J. Robbins (Fred Hutchinson Cancer Research Center), and Garland E. Allen (Washington University in St. Louis) for assistance in preparing the first edition. I am indebted to Peter and Rosemary Grant for figure-use permission. I also want to thank my post-doctoral mentor, Colin Hughes (now of Florida Atlantic University). I send a hearty "War Eagle!" to my friends, former students, and colleagues from Auburn University, especially Mike and Marie Wooten, Sharon Roberts, and Shreekumar Pulai.

My deepest gratitude goes to my husband, Douglas, who hikes with me, makes me laugh, and keeps my perspective balanced. Finally, I thank my mom and dad for love, support, prayers, and gumbo.

Publisher's Acknowledgments

We're proud of this book; please send us your comments at http://dummies.custhelp.com. For other comments, please contact our Customer Care Department within the U.S. at 877-762-2974, outside the U.S. at 317-572-3993, or fax 317-572-4002.

Some of the people who helped bring this book to market include the following:

Acquisitions, Editorial, and Media Development

Project Editor: Elizabeth Rea
  (Previous Edition: Mike Baker)

Acquisitions Editor: Michael Lewis
  (Previous Edition: Stacy Kennedy)

Copy Editor: Todd Lothery
  (Previous Edition: Elizabeth Rea)

Assistant Editor: Erin Calligan Mooney

Senior Editorial Assistant: David Lutton

Technical Editor: Lisa J. Cushman

Editorial Manager: Michelle Hacker

Editorial Assistant: Jennette ElNaggar

Cover Photos: iStock

Cartoons: Rich Tennant
  (www.the5thwave.com)

Composition Services

Project Coordinator: Katherine Crocker

Layout and Graphics: Ashley Chamberlain, Joyce Haughey

Proofreaders: Melissa Cossell, Leeann Harney

Indexer: Slivoskey Indexing Services

Publishing and Editorial for Consumer Dummies

  Kathleen Nebenhaus, Vice President and Executive Publisher

  David Palmer, Associate Publisher

  Kristin Ferguson-Wagstaffe, Product Development Director

Publishing for Technology Dummies

  Andy Cummings, Vice President and Publisher

Composition Services

  Debbie Stailey, Director of Composition Services

Contents at a Glance

Table of Contents

Introduction

Genetics affects all living things. Although sometimes complicated and always diverse, all genetics comes down to basic principles of *heredity* — how traits are passed from one generation to the next — and how DNA is put together. As a science, genetics is a fast growing field because of its untapped potential — for good and for bad. Despite its complexity, genetics can be surprisingly accessible. Genetics is a bit like peeking behind a movie's special effects to find a deceptively simple and elegant system running the whole show.

About This Book

Genetics For Dummies, 2nd Edition, is an overview of the entire field of genetics. My goal is to explain every topic so that anyone, even someone without any genetics background at all, can follow the subject and understand how it works. As in the first edition, I include many examples from the frontiers of research. I also make sure that the book has detailed coverage of some of the hottest topics that you hear about in the news: cloning, gene therapy, and forensics. And I address the practical side of genetics: how it affects your health and the world around you. In short, this book is designed to be a solid introduction to genetics basics and to provide some details on the subject.

Genetics is a fast-paced field; new discoveries are coming out all the time. You can use this book to help you get through your genetics course or for self-guided study. *Genetics For Dummies,* 2nd Edition, provides enough information for you to get a handle on the latest press coverage, understand the genetics jargon that mystery writers like to toss around, and translate information imparted to you by medical professionals. The book is filled with stories of key discoveries and "wow" developments. Although I try to keep things light and inject some humor when possible, at the same time, I make every effort to be sensitive to whatever your circumstances may be.

This book is a great guide if you know nothing at all about genetics. If you already have some background, then you're set to dive into the details of the subject and expand your horizons.

Conventions Used in This Book

I teach genetics in a university. It would be very easy for me to use specialized language that you'd need a translator to understand, but what fun would that be? Throughout this book, I avoid jargon as much as possible, but at the same time, I use and carefully define terms that scientists actually use. After all, it may be important for you to understand some of these multisyllabic jawbreakers in the course of your studies or your, or a loved one's, medical treatment.

To help you navigate through this book, I also use the following typographical conventions:

- ✔ I use _italic_ for emphasis and to highlight new words or terms that I define in the text.
- ✔ I use **boldface** to indicate keywords in bulleted lists or the action parts of numbered steps.
- ✔ I use `monofont` for Web sites and e-mail addresses. Note that some Web addresses may break across two lines of text. In such cases, I haven't inserted any hyphens to indicate a break. So if you type exactly what you see — pretending that the line break doesn't exist — you can get to your Web destination.

What You're Not to Read

Anytime you see a Technical Stuff icon (see "Icons Used in This Book" later in this introduction), you can cruise past the information it's attached to without missing a key explanation. For the serious reader (or a student intent on earning a high score), the technical bits add depth and detail to the book. You also have permission to skip the shaded gray boxes known as sidebars. Doing so doesn't affect your understanding of the subject at hand, but I pull together lots of amazing details in these boxes — from how aging affects your DNA (and vice versa) to how genetics affects your food — so I'm guessing (or at least hoping!) that the sidebars will grab your attention more often than not.

Foolish Assumptions

It's a privilege to be your guide into the amazing world of genetics. Given this responsibility, you were in my thoughts often while I was writing this book. Here's how I imagine you, my reader:

- ✔ You're a student in a genetics or biology class.
- ✔ You're curious to understand more about the science you hear reported in the news.
- ✔ You're an expectant or new parent or a family member who's struggling to come to terms with what doctors have told you.
- ✔ You're affected by cancer or some hereditary disease, wondering what it means for you and your family.

If any of these descriptions fit, you've come to the right place.

How This Book Is Organized

I designed this book to cover background material in the first two parts and then all the applications in the rest of the book. I think you'll find it quite accessible.

Part 1: The Lowdown on Genetics: Just the Basics

This part explains how trait inheritance works. The first chapter gives you a handle on how genetic information gets divvied up during cell division; these events provide the foundation for just about everything else that has to do with genetics. From there, I explain simple inheritance of one gene and then move on to more complex forms of inheritance. This part ends with an explanation of how sex works — that is, how genetics determines maleness or femaleness and how sex affects how your genes work. (If you're wondering how sex *really* works, check out *Sex For Dummies,* coauthored by Dr. Ruth.)

Part 11: DNA: The Genetic Material

This part covers what's sometimes called *molecular genetics*. But don't let the word "molecular" scare you off. It's the nitty-gritty details all right, but broken down so that you can easily follow along. I track the progress of how your genes work from start to finish: how your DNA is put together, how it gets copied, and how the building plans for your body are encoded in the double helix. To help you understand how scientists explore the secrets stored in your DNA, I cover how DNA is sequenced. In the process, I relate the fascinating story behind the Human Genome Project.

Part III: Genetics and Your Health

Part III is intended to help you see how genetics affects your health and well-being. I cover the subjects of genetic counseling, inherited diseases, genetics and cancer, and chromosome disorders such as Down syndrome. I also include a chapter on gene therapy, a practice that may hold the key to cures or treatments for many of the disorders I describe in this part of the book.

Part IV: Genetics and Your World

This part explains the broader impact of genetics and covers some hot topics that are often in the news. I explain how various technologies work and highlight both the possibilities and the perils of each. I delve into population genetics (of both humans, past and present, and endangered animal species), evolution, DNA and forensics, genetically modified plants and animals, cloning, and the issue of ethics, which is raised on a daily basis as scientists push the boundaries of the possible with cutting-edge technology.

Part V: The Part of Tens

In Part V, you get my lists of ten milestone events and important people that have shaped genetics history, ten of the next big things in the field, and ten "believe it or not" stories that can provide you with more insights on the issues found elsewhere in the book that interest you.

Icons Used in This Book

All *For Dummies* books use icons to help readers keep track of what's what. Here's a rundown of the icons I use in this book and what they all mean.

This icon flags information that's critical to your understanding or that's particularly important to keep in mind.

Points in the text where I provide added insight on how to get a better handle on a concept are found here. I draw on my teaching experience for these tips and alert you to other sources of information you can check out.

These details are useful but not necessary to know. If you're a student, though, these sections may be especially important to you.

This icon points out stories about the people behind the science and accounts of how discoveries came about.

This fine piece of art alerts you to recent applications of genetics in the field or in the lab.

Where to Go from Here

With *Genetics For Dummies,* 2nd Edition, you can start anywhere, in any chapter, and get a handle on what you're interested in right away. I make generous use of cross-references throughout the book to help you get background details that you may have skipped earlier. The table of contents and index can point you to specific topics in a hurry, or you can just start at the beginning and work your way straight through. If you read the book from front to back, you'll get a short course in genetics in the style and order that it's often taught in colleges and universities — Mendel first and DNA second.

Part I
The Lowdown on Genetics: Just the Basics

The 5th Wave   By Rich Tennant

"The results of our genetic testing have come in, and it appears you may share some DNA with the people of—get this—Easter Island."

In this part . . .

First and foremost, genetics is concerned with how traits are inherited. The process of cell division is central to how chromosomes are divvyed up among offspring. When genes are passed on, some are assertive and dominant while others are shy and recessive. The study of how different traits are inherited and expressed is called *Mendelian genetics*.

Genetics also determines your sex (as in maleness or femaleness), and your sex influences how certain traits are expressed. In this part, I explain what genetics is and what it's used for, how cells divide, and how traits are passed from parents to offspring.

Chapter 1

What Genetics Is and Why You Need to Know Some

In This Chapter

▶ Defining the subject of genetics and its various subdivisions

▶ Observing the day-to-day activities in a genetics lab

▶ Getting the scoop on career opportunities in genetics

Welcome to the complex and fascinating world of genetics. Genetics is all about physical traits and the DNA code that supplies the building plans for any organism. This chapter explains what the field of genetics is and what geneticists do. You get an introduction to the big picture and a glimpse at some of the details found in other chapters of this book.

What Is Genetics?

Genetics is the field of science that examines how traits are passed from one generation to the next. Simply put, genetics affects *everything* about *every* living thing on earth. An organism's *genes,* snippets of DNA that are the fundamental units of heredity, control how the organism looks, behaves, and reproduces. Because all biology depends on genes, understanding genetics as a foundation for all other life sciences, including agriculture and medicine, is critical.

From a historical point of view, genetics is still a young science. The principles that govern inheritance of traits by one generation from another were described (and promptly lost) less than 150 years ago. Around the turn of the 20th century, the laws of inheritance were rediscovered, an event that transformed biology forever. But even then, the importance of the star of the genetics show, DNA, wasn't really understood until the 1950s. Now, technology is helping geneticists push the envelope of knowledge every day.

Genetics is generally divided into four major subdivisions:

- ✓ **Classical, or Mendelian, genetics:** A discipline that describes how physical characteristics (traits) are passed along from one generation to another.
- ✓ **Molecular genetics:** The study of the chemical and physical structures of DNA, its close cousin RNA, and proteins. Molecular genetics also covers how genes do their jobs.
- ✓ **Population genetics:** A division of genetics that looks at the genetic makeup of larger groups.
- ✓ **Quantitative genetics:** A highly mathematical field that examines the statistical relationships between genes and the traits they encode.

In the academic world, many genetics courses begin with classical genetics and proceed through molecular genetics, with a nod to population, evolutionary, or quantitative genetics. This book follows the same path, because each division of knowledge builds on the one before it. That said, it's perfectly okay, and very easy, to jump around among disciplines. No matter how you take on reading this book, I provide lots of cross references to help you stay on track.

Classical genetics: Transmitting traits from generation to generation

At its heart, *classical genetics* is the genetics of individuals and their families. It focuses mostly on studying physical traits, or *phenotypes,* as a stand-in for the genes that control appearance.

Gregor Mendel, a humble monk and part-time scientist, founded the entire discipline of genetics. Mendel was a gardener with an insatiable curiosity to go along with his green thumb. His observations may have been simple, but his conclusions were jaw-droppingly elegant. This man had no access to technology, computers, or a pocket calculator, yet he determined, with keen accuracy, exactly how inheritance works.

Classical genetics is sometimes referred to as:

- ✓ **Mendelian genetics:** You start a new scientific discipline, and it gets named after you. Seems fair.
- ✓ **Transmission genetics:** This term refers to the fact that classical genetics describes how traits are passed on, or *transmitted,* by parents to their offspring.

No matter what you call it, classical genetics includes the study of cells and chromosomes (which I delve into in Chapter 2). Cell division is the machine that drives inheritance, but you don't have to understand combustion engines to drive a car, right? Likewise, you can dive straight into simple inheritance (see Chapter 3) and work up to more complicated forms of inheritance (in Chapter 4) without knowing anything whatsoever about cell division. (Mendel didn't know anything about chromosomes and cells when he figured this whole thing out, by the way.)

The genetics of sex and reproduction are also part of classical genetics. Various combinations of genes and chromosomes (strands of DNA) determine sex, as in maleness and femaleness. But the subject of sex gets even more complicated (and interesting): The environment plays a role in determining the sex of some organisms (like crocodiles and turtles), and other organisms can even change sex with a change of address. If I've piqued your interest, you can find out all the slightly kinky details in Chapter 5.

Classical genetics provides the framework for many subdisciplines. Genetic counseling (which I cover in Chapter 12) depends heavily on understanding patterns of inheritance to interpret people's medical histories from a genetics perspective. The study of chromosome disorders such as Down syndrome (see Chapter 15) relies on cell biology and an understanding of what happens during cell division. Forensics (see Chapter 18) also uses Mendelian genetics to determine paternity and to work out who's who with DNA fingerprinting.

Molecular genetics: DNA and the chemistry of genes

Classical genetics concentrates on studying outward appearances, but the study of actual genes falls under the heady title of *molecular genetics*. The area of operations for molecular genetics includes all the machinery that runs cells and manufactures the structures called for by the plans found in genes. The focus of molecular genetics includes the physical and chemical structures of the double helix, DNA, which I break down in all its glory in Chapter 6. The messages hidden in your DNA (your genes) constitute the building instructions for your appearance and everything else about you — from how your muscles function and how your eyes blink to your blood type, your susceptibility to particular diseases, and everything in between.

Your genes are expressed through a complex system of interactions that begins with copying DNA's messages into a somewhat temporary form called RNA (see Chapter 9). RNA carries the DNA message through the process of *translation* (covered in Chapter 10), which, in essence, is like taking a blueprint to a factory to guide the manufacturing process. Where your genes are concerned, the factory makes the proteins (from the RNA blueprint) that get folded in complex ways to make you.

The study of *gene expression* (how genes get turned on and off; flip to Chapter 11) and how the genetic code works at the levels of DNA and RNA are considered parts of molecular genetics. Research on the causes of cancer and the hunt for a cure (which I address in Chapter 14) focuses on the molecular side of things, because changes (referred to as *mutations*) occur at the chemical level of DNA (see Chapter 13 for coverage of mutations). Gene therapy (see Chapter 16), genetic engineering (see Chapter 19), and cloning (see Chapter 20) are all subdisciplines of molecular genetics.

Population genetics: Genetics of groups

Much to the chagrin of many undergrads, genetics is surprisingly mathematical. One area in which calculations are used to describe what goes on genetically is population genetics.

If you take Mendelian genetics and examine the inheritance patterns of many different individuals who have something in common, like geographic location, then you have population genetics. *Population genetics* is the study of the genetic diversity of a subset of a particular species (for details, jump to Chapter 17). In essence, it's a search for patterns that help describe the genetic signature of a particular group, such as the consequences of travel, isolation (from other populations), mating choices, geography, and behavior.

Population genetics helps scientists understand how the collective genetic diversity of a population influences the health of individuals within the population. For example, cheetahs are lanky cats; they're the speed demons of Africa. Population genetics has revealed that all cheetahs are very, very genetically similar; in fact, they're so similar that a skin graft from one cheetah would be accepted by any other cheetah. Because the genetic diversity of cheetahs is so low, conservation biologists fear that a disease could sweep through the population and kill off all the individuals of the species. It's possible that no animals would be resistant to the disease, and therefore, none would survive, leading to the extinction of this amazing predator.

Describing the genetics of populations from a mathematical standpoint is critical to forensics (see Chapter 18). To pinpoint the uniqueness of one DNA fingerprint, geneticists have to sample the genetic fingerprints of many individuals and decide how common or rare a particular pattern may be. Medicine also uses population genetics to determine how common particular mutations are and to develop new medicines to treat disease. For details on mutations, flip to Chapter 13; see Chapter 21 for information on genetics and the development of new medicines. Also, *evolutionary genetics,* or how traits change over time, is new to this edition; I cover the subject in Chapter 17.

Quantitative genetics: Getting a handle on heredity

Quantitative genetics examines traits that vary in subtle ways and relates those traits to the underlying genetics of an organism. A combination of whole suites of genes and environmental factors controls characteristics like retrieving ability in dogs, egg size or number in birds, and running speed in humans. Mathematical in nature, quantitative genetics takes a rather complex statistical approach to estimate how much variation in a particular trait is due to the environment and how much is actually genetic.

One application of quantitative genetics is determining how heritable a particular trait is. This measure allows scientists to make predictions about how offspring will turn out based on characteristics of the parent organisms. Heritability gives some indication of how much a characteristic (like seed production) can change when selective breeding (or, in evolutionary time, natural selection) is applied.

Living the Life of a Geneticist

Daily life for a geneticist can include working in the lab, teaching in the classroom, and interacting with patients and their families. In this section, you discover what a typical genetics lab is like and get a rundown of a variety of career paths in the genetics field.

Exploring a genetics lab

A genetics lab is a busy, noisy place. It's full of equipment and supplies and researchers toiling away at their workstations (called *lab benches,* even though the bench is really just a raised, flat surface that's conducive to working while standing up). Depending on the lab, you may see people looking very official in white lab coats or researchers dressed more casually in jeans and T-shirts. Every lab contains some or all of the following:

- ✔ Disposable gloves to protect workers from chemical exposure and to protect DNA and other materials from contamination.
- ✔ Pipettes (for measuring even the tiniest droplets of liquids with extreme accuracy), glassware (for liquid measurement and storage), and vials and tubes (for chemical reactions).
- ✔ Electronic balances for making super-precise measurements of mass.
- ✔ Chemicals and ultrapure water.

✔ A refrigerator (set at 40 degrees Fahrenheit), a freezer (at –4 degrees), and an ultracold freezer (at –112 degrees) for storing samples.

Repeated freezing and thawing causes DNA to break into tiny pieces, which destroys it. For that reason, freezers used in genetics labs aren't frost-free, because the temperature inside a frost-free freezer cycles up and down to melt any ice that forms.

✔ Centrifuges for separating substances from each other. Given that different substances have different densities, centrifuges spin at extremely high speeds to force materials to separate so that researchers can handle them individually.

✔ Incubators for growing bacteria under controlled conditions. Researchers often use bacteria for experimental tests of how genes work.

✔ Autoclaves for sterilizing glassware and other equipment using extreme heat and pressure to kill bacteria and viruses.

✔ Complex pieces of equipment such as thermocyclers (used for PCR; see Chapter 18) and DNA sequencers (see Chapter 8).

✔ Lab notebooks for recording every step of every reaction or experiment in nauseating detail. Geneticists must fully replicate (run over and over) every experiment to make sure the results are valid. The lab notebook is also a legal document that can be used in court cases, so precision and completeness are musts.

✔ Desktop computers packed with software for analyzing results and for connecting via the Internet to vast databases packed with genetic information (flip to the end of this chapter for the addresses of some useful Web sites).

Researchers in the lab use the various pieces of equipment and supplies from the preceding list to conduct experiments and run chemical reactions. Some of the common activities that occur in the genetics lab include

✔ Separating DNA from the rest of a cell's contents (see Chapter 6).

✔ Measuring the purity of a DNA sample and determining how much DNA (by weight) is present.

✔ Mixing chemicals that are used in reactions and experiments designed to analyze DNA samples.

✔ Growing special strains of bacteria and viruses to aid in examining short stretches of DNA (see Chapter 16).

✔ Using DNA sequencing (which I cover in Chapter 8) to learn the order of bases that compose a DNA strand (which I explain in Chapter 6).

✔ Setting up polymerase chain reactions, or PCR (see Chapter 18), a powerful process that allows scientists to analyze even very tiny amounts of DNA.

> ✔ Analyzing the results of DNA sequencing by comparing sequences from many different organisms (you can find this information in a massive, publicly available database — see the end of this chapter).
>
> ✔ Comparing DNA fingerprints from several individuals to identify perpetrators or to assign paternity (see Chapter 18).
>
> ✔ Holding weekly or daily meetings where everyone in the lab comes together to discuss results or plan new experiments.

Sorting through jobs in genetics

Whole teams of people contribute to the study of genetics. The following are just a few job descriptions for you to mull over if you're considering a career in genetics.

Lab tech

Lab technicians handle most of the day-to-day work in the lab. The tech mixes chemicals for everyone else in the lab to use in experiments. Techs usually prepare the right sorts of materials to grow bacteria (which are used as carriers for DNA; see Chapter 16), set up the bacterial cultures, and monitor their growth. Techs are also usually responsible for keeping all the necessary supplies straight and washing the glassware — not a glamorous job but a necessary one, because labs use tons of glass beakers and flasks that have to be cleaned.

When it comes to actual experiments, lab technicians are responsible for separating the DNA from the rest of the tissue around it and testing it for purity (to make sure no contaminants, like proteins, are present). Using a rather complicated machine with a strong laser, the tech can also measure exactly how much DNA is present. When a sufficiently pure sample of DNA is obtained, techs may analyze the DNA in greater detail (with PCR or sequencing reactions).

The educational background needed to be a lab tech varies with the amount of responsibility a particular position demands. Most techs have a minimum of a bachelor's degree in biology or some related field and need some background in microbiology to understand and carry out the techniques of handling bacteria safely and without contaminating cultures. And all techs must be good record-keepers, because every single activity in the lab must be documented in writing in the lab notebook.

Graduate student and post-doc

At most universities, genetics labs are full of *graduate students* working on either master's degrees or PhDs. In some labs, these students may be carrying out their own, independent research. On the other hand, many labs focus their work on a specific problem, like some specialized approach to studying cancer, and every student in that sort of lab works on some aspect of what his or her professor studies. Graduate students do a lot of the same things that lab techs do (see the preceding section), as well as design experiments, carry out those experiments, analyze the results, and then work to figure out what the results mean. Then, the graduate student writes a long document (called a *thesis* or *dissertation*) to describe what was done, what it means, and how it fits in with other people's research on the subject. While working in the lab, grad students take classes and are subjected to grueling exams (trust me on the grueling part).

All graduate students must hold a bachelor's degree. Performance on the standardized GRE (Graduate Record Exam) determines eligibility for admission to master's programs and may be used for selection for fellowships and awards.

If you're going to be staring down this test in the near future, you may want to get a leg up by checking out *The GRE Test For Dummies,* by Suzee Vlk, Michelle Rose Gilman, and Veronica Saydak (Wiley).

In general, it takes two or three years to earn a master's degree. A doctorate (denoted by *PhD*) usually requires anywhere from four to seven years of education beyond the bachelor's level.

After graduating with a PhD, a geneticist-in-training may need to get more experience before hitting the job market. Positions that provide such experience are collectively referred to as *post-docs* (post-doctoral fellows). A person holding a post-doc position is usually much more independent than a grad student when it comes to research. The post-doc often works to learn new techniques or to acquire a specialty before moving on to a position as a professor or a research scientist.

Research scientist

Research scientists work in private industries, designing experiments and directing the activities of lab techs. All sorts of industries employ research scientists, including

- Pharmaceutical companies, to conduct investigations on how drugs affect gene expression (see Chapter 11) and to develop new treatments such as gene therapy (see Chapter 16).

- Forensics labs, to analyze DNA found at crime scenes and to compare DNA fingerprints (see Chapter 18).

- Companies that analyze information generated by genome projects (human and others; see Chapter 11).

- Companies that support the work of other genetics labs by designing and marketing products used in research, such as kits used to run DNA fingerprints.

A research scientist usually holds a master's degree or a PhD. With only a bachelor's degree, several years of experience as a lab tech may suffice. Research scientists have to be able to design experiments and analyze results using statistics. Good record-keeping and strong communication skills (especially in writing) are musts. Most research scientists also have to be capable of managing and supervising people. In addition, financial responsibilities may include keeping up with expenditures, ordering equipment and supplies, and wrangling salaries of other personnel.

College or university professor

Professors do everything that research scientists do with the added responsibilities of teaching courses, writing proposals to get funds to support research, and writing papers on their research results for publication in reputable, peer-reviewed journals. Professors also supervise the lab techs, graduate students, and post-docs who work in their labs, which entails designing research projects and then ensuring that the projects are done correctly in the right amount of time (and under budget!).

Small schools may require a professor to teach as many as three courses every semester. Upper-tier institutions (think Big Ten or Ivy League) may require only one course of instruction per year. Genetics professors teach the basics as well as advanced and specialty courses like recombinant DNA (see Chapter 16) and population genetics (see Chapter 17).

To qualify for a professorship, universities require a minimum of a PhD, and most require additional post-doctoral experience. Job candidates must have already published research results to demonstrate the ability to do relevant research. Most universities also look for evidence that the professor-to-be will be successful at getting grants, which means the candidate must usually land a grant before getting a job.

Genetic counselor

Genetic counselors work with medical personnel to interpret the medical histories of patients and their families. The counselor usually works directly with the patient to assemble all the information into a family tree (see Chapter 12) and then looks for patterns to determine which traits may be hereditary. Counselors can also tell which diseases a patient is most likely to inherit. Genetic counselors are trained to conduct careful and thorough interviews to make sure that no information is missed or left out.

Genetic counselors usually hold a master's degree. Training includes many hours working with patients to hone interview and analysis skills (under the close supervision of experienced professionals, of course). The position requires excellent record-keeping skills and strict attention to detail. Genetic counselors also have to be good at interacting with all kinds of people, including research scientists and physicians. And the ability to communicate very well, both in writing and verbally, is a must.

The most essential skill of a genetic counselor is the ability to be nonjudgmental and nondirective. The counselor must be able to analyze a family history without bias or prejudice and inform the patient of his or her options without recommending any one course of action over another. Furthermore, the counselor must keep all information about his or her patients confidential, sharing information only with authorized personnel such as the person's own physician in order to protect the patient's privacy.

Great genetics Web sites to explore

The Internet is an unparalleled source of information about genetics. With just a few mouse clicks, you can find the latest discoveries and attend the best courses ever offered on the subject. Here's a quick sample.

✔ To see a great video that explains genetics and gives it a human face, check out "Cracking the Code of Life": www.pbs.org/wgbh/nova/genome/program.html.

✔ New discoveries are unveiled every day. To stay current, log on to www.sciencedaily.com/news/plants_animals/genetics/ and www.nature.com/ng/index.html.

✔ For students, http://learn.genetics.utah.edu/ can't be beat.

From the basics of heredity to virtual labs to cloning, it's all there in easy-to-grasp animations and language.

✔ Want to get all the details about genes and diseases? Start at www.ncbi.nlm.nih.gov/books/bv.fcgi?rid=gnd for the basics. You can find more advanced (and greatly detailed) information at Online Mendelian Inheritance in Man: www.ncbi.nlm.nih.gov/omim/.

✔ If you're interested in a career in genetics, the Genetics Society of America is ready to help: www.genetics-gsa.org/pages/careers_in_genetics.shtml.

Chapter 2

Basic Cell Biology

. . . .

In This Chapter

▶ Getting to know the cell

▶ Understanding chromosomes

▶ Exploring simple cell division

▶ Appreciating the complexities of meiosis

. . . .

Genetics and the study of how cells work are closely related. The process of passing genetic material from one generation to the next depends completely on how cells grow and divide. To reproduce, a simple organism such as bacteria or yeast simply copies its DNA (through a process called *replication,* which I cover in Chapter 7) and splits in two. But organisms that reproduce sexually go through a complicated dance that includes mixing and matching strands of DNA (a process called *recombination*) and then halving the amount of DNA for special sex cells, allowing completely new genetic combinations for their offspring. These amazing processes are part of what makes you unique. So come inside your cell — you need to be familiar with the processes of *mitosis* (cell division) and *meiosis* (the production of sex cells) to appreciate how genetics works.

Looking Around Your Cell

There are two basic kinds of organisms:

- ✔ **Prokaryotes:** Organisms whose cells lack a nucleus and therefore have DNA floating loosely in the liquid center of the cell
- ✔ **Eukaryotes:** Organisms that have a well-defined nucleus to house and protect the DNA

A *nucleus* is a compartment filled with DNA surrounded by a membrane.

The basic biologies of prokaryotes and eukaryotes are similar but not identical. Because all living things fall into these two groups, understanding the differences and similarities between cell types is important. In this section, I show you how to distinguish the two kinds of cells from each other, and you get a quick tour of the insides of cells — both with and without nuclei. Figure 2-1 shows you the structure of each type of cell.

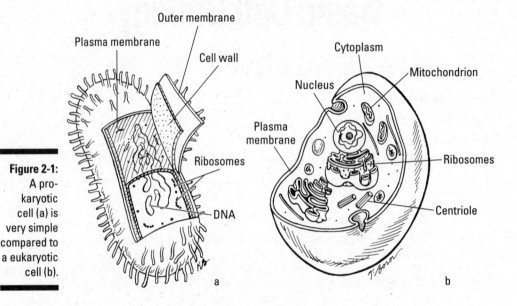

Figure 2-1: A prokaryotic cell (a) is very simple compared to a eukaryotic cell (b).

Cells without a nucleus

Scientists classify organisms composed of cells without nuclei as *prokaryotes,* which means "before nucleus." Prokaryotes are the most common forms of life on earth. You are, at this very moment, covered in and inhabited by millions of prokaryotic cells: bacteria. Much of your life and your body's processes depend on these arrangements; for example, the digestion going on in your intestines is partially powered by bacteria that break down the food you eat. Most of the bacteria in your body are completely harmless, but some species of bacteria can be vicious and deadly, causing rapidly transmitted diseases such as cholera.

All bacteria, regardless of temperament, are simple, one-celled, prokaryotic organisms. None has cell nuclei, and all are small cells with relatively small amounts of DNA (see Chapter 8 for more on the amounts of DNA different organisms possess).

The exterior of a prokaryotic cell is encapsulated by a *cell wall* that serves as the bacteria's only protection from the outside world. A *plasma membrane* (*membranes* are thin sheets or layers) regulates the exchange of nutrients, water, and gases that nourish the bacterial cell. DNA, usually in the form of a single, hoop-shaped piece, floats around inside the cell; segments of DNA like this one are called *chromosomes* (see the section "Examining the basics of chromosomes" later in the chapter). The liquid interior of the cell is called the *cytoplasm*. The cytoplasm provides a cushiony, watery home for the DNA and other cell machinery that carry out the business of living. Prokaryotes divide, and thus reproduce, by simple mitosis, which I cover in detail in the "Mitosis: Splitting Up" section later in the chapter.

Cells with a nucleus

Scientists classify organisms that have cells with nuclei as *eukaryotes,* which means "true nucleus." Eukaryotes range in complexity from simple, one-celled animals and plants to complex, multicellular organisms like you. Eukaryotic cells are fairly complicated and have numerous parts to keep track of (refer to Figure 2-1). Like prokaryotes, eukaryotic cells are held together by a *plasma membrane,* and sometimes a *cell wall* surrounds the membrane (plants, for example, have cell walls). But that's where the similarities end.

The most important feature of the eukaryotic cell is the *nucleus* — the membrane-surrounded compartment that houses the DNA that's divided into one or more chromosomes. The nucleus protects the DNA from damage during day-to-day living. Eukaryotic chromosomes are usually long, string-like segments of DNA instead of the hoop-shaped ones found in prokaryotes. Another hallmark of eukaryotes is the way the DNA is packaged: Eukaryotes usually have much larger amounts of DNA than prokaryotes, and to fit all that DNA into the tiny cell nucleus, it must be tightly wound around special proteins. (For all the details about DNA packaging for eukaryotes, flip to Chapter 6.)

Unlike prokaryotes, eukaryotes have all sorts of cell parts, called *organelles,* that help carry out the business of living. The organelles float around in the watery cytoplasm outside the nucleus. Two of the most important organelles are

- **Mitochondria:** The powerhouses of the eukaryotic cell, mitochondria pump out energy by converting glucose to ATP (adenosine triphosphate). ATP acts like a battery of sorts, storing energy until it's needed for day-to-day living. Both animals and plants have mitochondria.

- **Chloroplasts:** These organelles are unique to plants. They process the energy of sunlight into sugars that the plant mitochondria use to generate the energy that nourishes the living cells.

Eukaryotic cells are able to carry out behaviors that prokaryotes can't. For example, one-celled eukaryotes often have appendages, such as long tails (called *flagella*) or hair-like projections (called *cilia*), that work like hundreds of tiny paddles, helping them move around. Also, only eukaryotic cells are capable of ingesting fluids and particles for nutrition; prokaryotes must transport materials through their cell walls, a process that severely limits their dietary options.

In most multicellular eukaryotes, cells come in two basic varieties: body cells (called *somatic* cells) or sex cells. The two cell types have different functions and are produced in different ways.

Somatic cells

Somatic cells are produced by simple cell division called *mitosis* (see the section "Mitosis: Splitting Up" for details). Somatic cells of multicellular organisms like humans are differentiated into special cell types. Skin cells and muscle cells are both somatic cells, for instance, but if you were to examine your skin cells under a microscope and compare them with your muscle cells, you'd see that their structures are very different. The various cells that make up your body all have the same basic components (membrane, organelles, and so on), but the arrangements of the elements change from one cell type to the next so that they can carry out various jobs such as digestion (intestinal cells), energy storage (fat cells), or oxygen transport to your tissues (blood cells).

Sex cells

Sex cells are specialized cells used for reproduction. Only eukaryotic organisms engage in sexual reproduction, which I cover in detail at the end of this chapter in the section "Mommy, where did I come from?" *Sexual reproduction* combines genetic material from two organisms and requires special preparation in the form of a reduction in the amount of genetic material allocated to sex cells — a process called *meiosis* (see "Meiosis: Making Cells for Reproduction" later in the chapter for an explanation). In humans, the two types of sex cells are eggs and sperm.

Examining the basics of chromosomes

Chromosomes are threadlike strands composed of DNA. To pass genetic traits from one generation to the next, the chromosomes must be copied (see Chapter 7), and then the copies must be divvied up. Most prokaryotes have only one circular chromosome that, when copied, is passed on to the *daughter cells* (new cells created by cell division) during mitosis. Eukaryotes have more complex problems to solve (like divvying up half of the chromosomes to make sex cells), and their chromosomes behave differently during mitosis and meiosis. Additionally, various scientific terms describe the anatomy, shapes, number of copies, and situations of eukaryotic chromosomes. This section gets into the intricacies of chromosomes in eukaryotic cells because they're so complex.

Counting out chromosome numbers

Each eukaryotic organism has a specific number of chromosomes per cell — ranging from one to many. For example, humans have 46 total chromosomes. These chromosomes come in two varieties:

✔ **Sex chromosomes:** These chromosomes determine gender. Human cells contain two sex chromosomes. If you're female, you have two X chromosomes, and if you're male, you have an X and a Y chromosome. (To find out more about how sex is determined by the X and Y chromosomes, flip to Chapter 5.)

✔ **Autosomal chromosomes:** *Autosomal* simply refers to non-sex chromosomes. Sticking with the human example, if you do the math, you can see that humans have 44 autosomal chromosomes.

Ah, but there's more. In humans, chromosomes come in pairs. That means you have 22 pairs of uniquely shaped autosomal chromosomes plus 1 pair of sex chromosomes for a total of 23 chromosome pairs. Your autosomal chromosomes are identified by numbers — 1 through 22. So you have two chromosome 1s, two 2s, and so on. Figure 2-2 shows you how all human chromosomes are divided into pairs and numbered. (A *karyotype* like the one pictured in Figure 2-2 is one way chromosomes are examined; you discover more about karyotyping in Chapter 15.)

When chromosomes are sorted into pairs, the individual chromosomes in each pair are considered *homologous,* meaning that the paired chromosomes are identical to one another according to which genes they carry. In addition, your homologous chromosomes are identical in shape and size. These pairs of chromosomes are sometimes referred to as *homologs* for short.

Chromosome numbers can be a bit confusing. Humans are *diploid,* meaning we have two copies of each chromosome. Some organisms (like bees and wasps) have only one set of chromosomes (cells with one set of chromosomes are called *haploid*); others have three, four, or as many as sixteen copies of each chromosome! The number of chromosome sets held by a particular organism is called the *ploidy.* For more on chromosome numbers, see Chapter 15.

The total number of chromosomes doesn't tell you what the ploidy of an organism is. For that reason, the number of chromosomes an organism has is often listed as some multiple of *n*. A single set of chromosomes referred to by the *n* is the haploid number. Humans are $2n = 46$ (indicating that humans are diploid and their total number of chromosomes is 46). Human sex cells such as eggs or sperm are haploid (see "Mommy, where did I come from?" later in this chapter).

Geneticists believe that the homologous pairs of chromosomes in humans started as one set (that is, *haploid*), and the entire set was duplicated at some point in some distant ancestor, many millions of years ago.

Examining chromosome anatomy

Chromosomes are often depicted in stick-like forms, like those you see in Figure 2-3. Chromosomes don't look like sticks, though. In fact, most of the time they're loose and string-like. Chromosomes only take on this distinctive shape and form when cell division is about to take place (during metaphase of meiosis or mitosis). They're often drawn this way so that the special characteristics of eukaryotic chromosomes are easier to see. Figure 2-3 points out the important features of eukaryotic chromosomes.

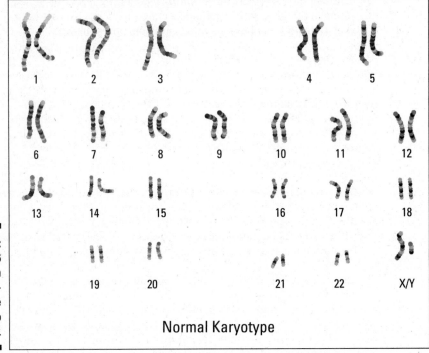

Figure 2-2:
The 46 human chromosomes are divided into 23 pairs.

Normal Karyotype

The part of the chromosome that appears pinched (in Figure 2-3, located in the middle of the chromosomes) is called the *centromere*. The placement of the centromere (whether it's closer to the top, middle, or bottom of the chromosome; see Figure 2-4) is what gives each chromosome its unique shape. The ends of the chromosomes are called *telomeres*. Telomeres are made of densely packed DNA and serve to protect the DNA message that the chromosome carries. (Flip to Chapter 23 for more about telomeres and how they may affect the process of aging.)

The differences in shapes and sizes of chromosomes are easy to see, but the most important differences between chromosomes are hidden deep inside the DNA. Chromosomes carry *genes* — sections of DNA that make up the building plans for physical traits. The genes tell the body how, when, and where to make all the structures that are necessary for the processes of living (for more on how genes work, flip to Chapter 11). Each pair of homologous chromosomes carries the same — but not necessarily identical — genes. For example, both chromosomes of a particular homologous pair may contain the gene for hair color, but one can be a "brown hair" version of the gene — alternative versions of genes are called *alleles* (refer to Figure 2-3) — and the other can be a "blond hair" allele.

Any given gene can have one or more alleles. In Figure 2-3, one chromosome carries the allele *A* while its homolog carries the allele *a* (the relative size of an allele is very small; the alleles are large here so you can see them). The alleles code for the different physical traits *(phenotypes)* you see in animals and plants, like hair color or flower shape. You can find out more about how alleles affect phenotype in Chapter 3.

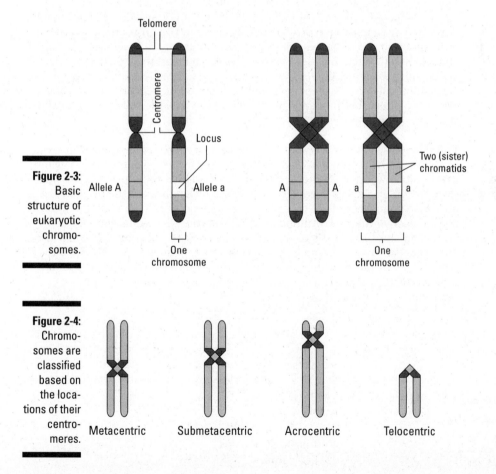

Figure 2-3: Basic structure of eukaryotic chromosomes.

Figure 2-4: Chromosomes are classified based on the locations of their centromeres.

Each point along the chromosome is called a *locus* (Latin for "place"). The plural of locus is *loci* (pronounced *low*-sigh). Most of the phenotypes that you see are produced by multiple genes (that is, genes occurring at different loci and often on different chromosomes) acting together. For instance, human eye color is determined by at least three genes that reside on two different chromosomes. You can find out more about how genes are arranged along chromosomes in Chapter 15.

Mitosis: Splitting Up

Most cells have simple lifestyles: They grow, divide, and eventually die. Figure 2-5 illustrates the basic life cycle of a typical somatic, or body, cell.

The *cell cycle* (the stages a cell goes through from one division to another) is tightly regulated; some cells divide all the time, and others never divide at all. Your body uses mitosis to provide new cells when you grow and to replace cells that wear out or become damaged from injury. Talk about multitasking — you're going through mitosis right now, while you read this book! Some cells divide only part of the time, when new cells are needed to handle certain jobs like fighting infection. Cancer cells, on the other hand, get carried away and divide too often. (In Chapter 14, you can find out how the cell cycle is regulated and what happens when it goes awry.)

The cell cycle includes *mitosis* — the process of reproducing the cell nucleus by division. The result of each round of the cell cycle is a simple cell division that creates two identical new cells from one original cell. During mitosis, all DNA present in the cell is copied (see Chapter 7), and when the original cell divides, a complete collection of all the chromosomes (in humans, 23 pairs) goes to each of the two resulting cells. Prokaryotes and some simple eukaryotic organisms use mitosis to reproduce themselves. (More complex eukaryotic organisms use *meiosis* for sexual reproduction, in which each of the two sex cells sends only one copy of each chromosome into the eggs or sperm. You can read all about that in the section "Meiosis: Making Cells for Reproduction" later in this chapter.)

Figure 2-5:
The cell cycle: mitosis, cell division, and all points in between.

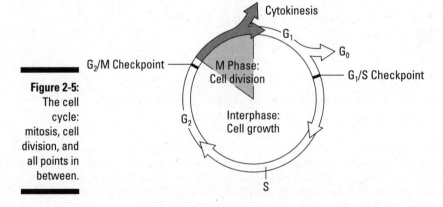

You should remember two important points about mitosis:

- **Mitosis produces two identical cells.** The new cells are identical to each other *and* to the cell that divided to create them.

- **Cells created by mitosis have exactly the same number of chromosomes as the original cell did.** If the original cell had 46 chromosomes, the new cells each have 46 chromosomes.

Mitosis is only one of the major phases in the cell cycle; the other is *interphase*. In the following sections, I guide you through the phases of the cell cycle and tell you exactly what happens during each one.

Step 1: Time to grow

Interphase is the part of the cell cycle during which the cell grows, copies its DNA, and prepares to divide. Interphase occurs in three stages: the G1 phase, the S phase, and the G2 phase.

G1 phase

When a cell begins life, such as the moment an egg is fertilized, the first thing that happens is the original cell starts to grow. This period of growth is called the *G1 phase* of interphase. Lots of things happen during G1: DNA supervises the work of the cell, *metabolism* (the exchange of oxygen and carbon dioxide) occurs, and cells breathe and "eat."

Some cells opt out of the cell cycle permanently, stop growing, and exit the process at G_0. Your brain cells, for example, have retired from the cell cycle. Mature red blood cells and muscle cells don't divide, either. In fact, human red blood cells have no nuclei and thus possess no DNA of their own.

If the cell in question plans to divide, though, it can't stay in G1 forever. Actively dividing cells go through the whole cell cycle every 24 hours or so. After a predetermined period of growth that lasts from a few minutes to several hours, the cell arrives at the first checkpoint (refer to Figure 2-5). When the cell passes the first checkpoint, there's no turning back.

Various proteins control when the cell moves from one phase of the cycle to the next. At the first checkpoint, proteins called *cyclins* and enzymes called *kinases* control the border between G1 and the next phase. Cyclins and kinases interact to cue up the various stages of the merry-go-round of cell division. Two particular chemicals, CDK (cyclin dependent kinase) and G1 cyclin, hook up to escort the cell over the border from G1 to S — the next phase.

S phase

S phase is the point at which the cell's DNA is replicated (here, *S* refers to *synthesis,* or copying, of the DNA). When the cell enters the S phase, activity around the chromosomes really steps up. All the chromosomes must be copied to make exact replicas that later are passed on to the newly formed daughter cells produced by cell division. DNA replication is a very complex process that gets full coverage in Chapter 7.

For now, all you need to know is that all the cell's chromosomes are copied during S, and the copies stay together as a unit (joined at the centromere; refer back to Figure 2-3) when the cell moves from S into G2 — the final step in interphase. The replicated chromosomes are called *sister chromatids* (refer to Figure 2-3), which are alike in every way. They carry the exact same copies of the exact same genes. During mitosis (or meiosis), the sister chromatids are divided up and sent to the daughter cells as part of the cell cycle.

G2 phase

The *G2 phase* leads up to cell division. It's the last phase before actual mitosis gets underway. G2, sometimes called *Gap 2,* gives the cell time to get bigger before splitting into two smaller cells. Another set of cyclins and CDK work together to push the cell through the second checkpoint located at the border between G2 and mitosis. (For details on the first checkpoint, jump back to the section "G1 phase.") As the cell grows, the chromosomes, now copied and hooked together as sister chromatids, stay together inside the cell nucleus. (The DNA is still "relaxed" at this point and hasn't yet taken on the fat, sausage-shaped appearance it assumes during mitosis.) After the cell crosses the G2/M checkpoint (refer to Figure 2-5), the business of mitosis formally gets underway.

Step 2: Divvying up the chromosomes

In the cell cycle, *mitosis* is the process of dividing up the newly copied chromosomes (that were created in interphase; see the preceding section) to make certain that the new cells each get a full set. Generally, mitosis is divided into four phases, which you can see in Figure 2-6 and read about in the following sections.

The phases of mitosis are a bit artificial, because the movement doesn't stop at each point; instead, the chromosomes cruise right from one phase to the next. But dividing the process into phases is useful for understanding how the chromosomes go from being all mixed together to neatly parting ways and getting into the proper, newly formed cells.

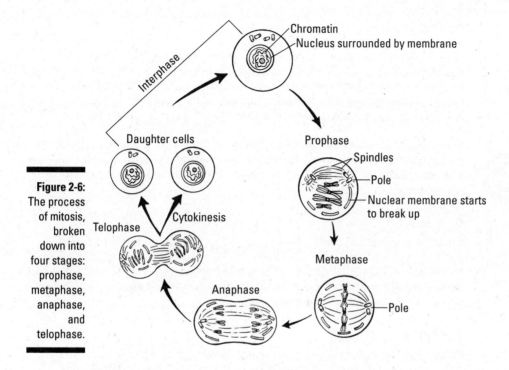

Figure 2-6:
The process
of mitosis,
broken
down into
four stages:
prophase,
metaphase,
anaphase,
and
telophase.

Prophase

During *prophase,* the chromosomes get very compact and condensed, taking on the familiar sausage shape. During interphase (see the "Step 1: Time to grow" section earlier in this chapter), the DNA that makes up the chromosomes is tightly wound around special proteins, sort of like string wrapped around beads. The whole "necklace" is wound tightly on itself to compress the enormous DNA molecules to sizes small enough to fit inside the cell nucleus. But even when coiled during interphase, the chromosomes are still so threadlike and tiny that they're essentially invisible. That changes during prophase, when the chromosomes become so densely packed that you can easily see them with an ordinary light microscope.

By the time they reach prophase, chromosomes have duplicated to form sister chromatids (refer to Figure 2-3). Sister chromatids of each chromosome are exact twin copies of each other. Each chromatid is actually a chromosome in its own right, but thinking of chromosomes as chromatids may help you keep all the players straight during the process of division.

As the chromosomes/chromatids condense, the cell nucleus starts breaking up, allowing the chromosomes to move freely across the cell as the process of cell division progresses.

Metaphase

Metaphase is the point when the chromosomes all line up in the center of the cell. After the nuclear membrane dissolves and prophase is complete, the chromosomes go from being a tangled mass to lining up in a more or less neat row in the center of the cell (refer to Figure 2-6). Threadlike strands called *spindles* grab each chromosome around its waist-like centromere. The spindles are attached to points on either side of the cell called *poles*.

Sometimes, scientists use geographic terms to describe the positions of chromosomes during metaphase: The chromosomes line up at the equator and are attached to the poles. This trick may help you better visualize the events of metaphase.

Anaphase

During *anaphase,* the sister chromatids are pulled apart, and the resulting halves migrate to opposite poles (refer to Figure 2-6). At this point, it's easy to see that the chromatids are actually chromosomes. Every sister chromatid gets split apart so that the cell that's about to be formed ends up with a full set of all the original cell's chromosomes.

Telophase

Finally, during *telophase,* nuclear membranes begin to form around the two sets of separated chromosomes (refer to Figure 2-6). The chromosomes begin to relax and take on their usual interphase form. The cell itself begins to divide as telophase comes to an end.

Step 3: The big divide

When mitosis is complete and new nuclei have formed, the cell divides into two smaller, identical cells. The division of one cell into two is called *cytokinesis* (*cyto* meaning "cell" and *kinesis* meaning "movement"). Technically, cytokinesis happens after metaphase is over and before interphase begins. Each new cell has a full set of chromosomes, just as the original cell did. All the organelles and cytoplasm present in the original cell are divided up to provide the new cell with all the machinery it needs for metabolism and growth. The new cells are now at interphase (specifically, the G1 stage) and are ready to begin the cell cycle again.

Meiosis: Making Cells for Reproduction

Meiosis is a cell division that includes reducing the chromosome number as preparation for sexual reproduction. Meiosis reduces the amount of DNA by half so that when fertilization occurs, each offspring gets a full set of chromosomes. As a result of meiosis, the cell goes from being diploid to being haploid. Or, to put it another way, the cell goes from being *2n* to being *n.* In humans, this means that the cells produced by meiosis (either eggs or sperm) have 23 chromosomes each — one copy of each of the homologous chromosomes. (See the section "Counting out chromosome numbers" earlier in this chapter for more information.)

Meiosis has many characteristics in common with mitosis. The stages go by similar names, and the chromosomes move around similarly, but the products of meiosis are completely different from those of mitosis. Whereas mitosis ends with two identical cells, meiosis produces *four* cells each with *half* the amount of DNA that the original cell contained. Furthermore, with meiosis, the homologous chromosomes go through a complex exchange of segments of DNA called *recombination.* Recombination is one of the most important aspects of meiosis and leads to genetic variation that allows each individual produced by sexual reproduction to be truly unique.

Meiosis goes through two rounds of division: meiosis I and the sequel, meiosis II. Figure 2-7 shows the progressing stages of both meiosis I and meiosis II. Unlike lots of movie sequels, the sequel in meiosis is really necessary. In both rounds of division, the chromosomes go through stages that resemble those in mitosis. However, the chromosomes undergo different actions in meiotic prophase, metaphase, anaphase, and telophase.

Students often get stuck on the phases of meiosis and miss its most important aspects: recombination and the division of the chromosomes. To prevent that sort of confusion, I don't break down meiosis by phases. Instead, I focus on the activities of the chromosomes themselves.

In meiosis I:

- ✔ The homologous pairs of chromosomes line up side by side and exchange parts. This is called *crossing-over* or *recombination,* and it occurs during prophase I.

- ✔ During metaphase I, the homologous chromosomes line up at the equator of the cell (called the *metaphase plate*), and homologs go to opposite poles during the first round of anaphase.

- ✔ The cell divides in telophase I, reducing the amount of genetic material by half, and enters a second round of division — meiosis II.

During meiosis II:

- ✔ The individual chromosomes (as sister chromatids) condense during prophase II and line up at the metaphase plates of both cells (metaphase II).

- ✔ The chromatids separate and go to opposite poles (anaphase II).

- ✔ The cells divide, resulting in a total of *four* daughter cells, each possessing *one* copy of each chromosome.

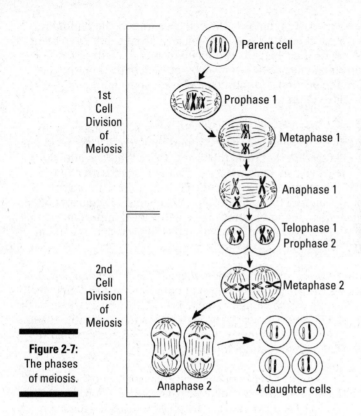

Figure 2-7:
The phases
of meiosis.

Meiosis part 1

Cells that undergo meiosis start in a phase similar to the interphase that pre-cedes mitosis. The cells grow in a G1 phase, undergo DNA replication during S, and prepare for division during G2. (To review what happens in each of these phases, flip back to the section "Step 1: Time to grow.") When meiosis is about to begin, the chromosomes condense. By the time meiotic inter-phase is complete, the chromosomes have been copied and are hitched up as sister chromatids, just as they would be in mitosis. Next up are the phases of meiosis I, which I profile in the sections that follow.

Find your partner

During prophase I (labeled "I" because it's in the first round of meiosis), the homologous chromosomes find each other. These homologous chromo-somes originally came from the mother and father of the individual whose cells are now undergoing meiosis. Thus, during meiosis, maternal and paternal chromosomes, as homologs, line up side by side. In Figure 2-2, you can see an entire set of 46 human chromosomes. Although the members of the pair seem identical, they're not. The homologous chromosomes have different combinations of alleles at the thousands of loci along each chromosome. (For more on alleles, jump to the section "Examining chromosome anatomy" earlier in this chapter.)

Recombining makes you unique

When the homologous chromosomes pair up in prophase I, the chromatids of the two homologs actually zip together, and the chromatids exchange parts of their arms. Enzymes cut the chromosomes into pieces and seal the newly combined strands back together in an action called *crossing-over.* When crossing-over is complete, the chromatids consist of part of their original DNA and part of their homolog's DNA. The loci don't get mixed up or turned around — the chromosome sequence stays in its original order. The only thing that's different is that the maternal and paternal chromosomes (as homologs) are now mixed together.

Figure 2-8 illustrates crossing-over in action. The figure shows one pair of homologous chromosomes and two loci. At both loci, the chromosomes have alternative forms of the genes. In other words, the alleles are different: Homolog one has *A* and *b,* and homolog two has *a* and *B.* When replication takes place, the sister chromatids are identical (because they're exact copies of each other). After crossing-over, the two sister chromatids have exchanged arms. Thus, each homolog has a sister chromatid that's different.

Partners divide

The recombined homologs line up at the metaphase equator of the cell (refer to Figure 2-7). The nuclear membrane begins to break down, and in a process similar to mitotic anaphase, spindle fibers grasp the homologous chromosomes by their centromeres and pull them to opposite sides of the cell.

At the end of the first phase of meiosis, the cell undergoes its first round of division (telophase 1, followed by cytokinesis 1). The newly divided cells each contain one set of chromosomes, the now partnerless homologs, still in the form of replicated sister chromatids.

When the homologs line up, maternal and paternal chromosomes pair up, but it's a tossup as to which side of the equator each one ends up on. Therefore, each pair of homologs divides independently of every other homologous pair. This is the basis of the principle of independent assortment, which I cover in Chapters 3 and 4.

Following telophase I, the cells enter an in-between round called *interkinesis* (which means "between movements"). The chromosomes relax and lose their fat, ready-for-metaphase appearance. Interkinesis is just a "resting" phase in preparation for the second round of meiosis.

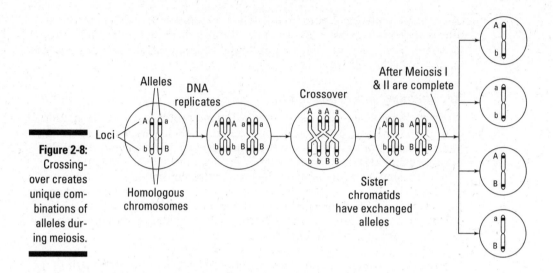

Figure 2-8:
Crossing-over creates unique combinations of alleles during meiosis.

Meiosis II: The sequel

Meiosis II is the second phase of cell division that produces the final product of meiosis: cells that contain only one copy of each chromosome. The chromosomes condense once more to their now-familiar fat, sausage shapes. Keep in mind that each cell has only a single set of chromosomes, which are still in the form of sister chromatids.

During metaphase II, the chromosomes line up along the equator of the cells, and spindle fibers attach at the centromeres. In anaphase II, the sister chromatids are pulled apart and move to opposite poles of their respective cell. The nuclear membranes form around the now single chromosomes (telophase II). Finally, cell division takes place. At the end of the process, each of the four cells contains one single set of chromosomes.

Mommy, where did I come from?

From gametogenesis, honey. Meiosis in humans (and in all animals that reproduce sexually) produces cells called *gametes*. Gametes come in the form of sperm (produced by males) or eggs (produced by females). When conditions are right, sperm and egg unite to create a new organism, which takes the form of a *zygote*. Figure 2-9 shows the process of *gametogenesis* (the production of gametes) in humans.

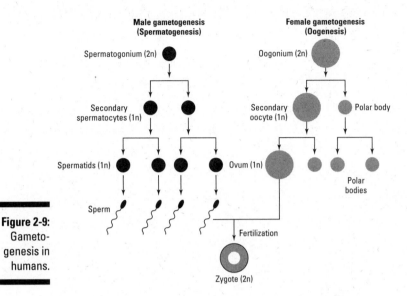

Figure 2-9:
Gameto-
genesis in
humans.

For human males, special cells in the male's sexual organs (testes) produce *spermatogonia*. Spermatogonia are *2n* — they contain a full diploid set of 46 chromosomes (see the earlier section "Counting out chromosome numbers"). After meiosis I, each single spermatogonium has divided into two cells called *secondary spermatocytes*. These spermatocytes contain only one copy of each homolog (as sister chromatids). After one more division (meiosis II), the *spermatids* that become sperm cells have one copy of each chromosome. Thus, sperm cells are haploid and contain 23 chromosomes. Because males have X and Y sex chromosomes, half their sperm (men produce literally millions) contain Xs and half contain Ys.

Human females produce eggs in much the same way that men produce sperm. Egg cells, which are produced by the ovaries, start as diploid *oogonia* (that is, *2n* = 46). The big difference between egg and sperm production is that at the end of meiosis II, only one mature, haploid (23 chromosomes) sex cell (as an egg) is produced instead of four (refer to Figure 2-9). The other three cells produced are called *polar bodies;* the polar bodies aren't actual egg cells and can't be fertilized to produce offspring.

Why does the female body produce one egg cell and three polar bodies? Egg cells need large amounts of cytoplasm to nourish the zygote in the period between fertilization and when the mother starts providing the growing embryo with nutrients and energy through the placenta. The easiest way to get enough cytoplasm into the egg when it needs it most is to put less cytoplasm into the other three cells produced in meiosis II.

Chapter 3

Visualize Peas: Discovering the Laws of Inheritance

*A*ll the physical traits of any living thing originate in that organism's genes. Look at the leaves of a tree or the color of your own eyes. How tall are you? What color is your dog's or cat's fur? Can you curl or fold your tongue? Got hair on the backs of your fingers? All that and much more came from genes passed down from parent to offspring. Even if you don't know much about how genes work or even what genes actually are, you've probably already thought about how physical traits can be inherited. Just think of the first thing most people say when they see a newborn baby: Who does he or she look most like, mommy or daddy?

The *laws of inheritance* — how traits are transmitted from one generation to the next (including dominant-recessive inheritance, segregation of alleles into gametes, and independent assortment of traits) — were discovered less than 200 years ago. In the early 1850s, Gregor Mendel, an Austrian monk with a love of gardening, looked at the physical world around him and, by simply growing peas, categorized the patterns of genetic inheritance that are still recognized today. In this chapter, you discover how Mendel's peas changed the way scientists view the world. If you skipped Chapter 2, don't worry — Mendel didn't know anything about mitosis or meiosis when he formulated the laws of inheritance.

Mendel's discoveries have an enormous impact on your life. If you're interested in how genetics affects your health (Part III), reading this chapter and getting a handle on the laws of inheritance will help you.

Gardening with Gregor Mendel

For centuries before Mendel planted his first pea plant, scholars and scientists argued about how inheritance of physical traits worked. It was obvious that *something* was passed from parent to offspring, because diseases and personality traits seemed to run in families. And farmers knew that by breeding plants and animals with certain physical features that they valued, they could create varieties that produced desirable products, like higher yielding maize, stronger horses, or hardier dogs. But just how inheritance worked and exactly what was passed from parent to child remained a mystery.

Enter the star of our gardening show, Gregor Mendel. Mendel was, by nature, a curious person. As he wandered around the gardens of the monastery where he lived in the mid-19th century, he noticed that his pea plants looked different from one another in a number of ways. Some were tall and others short. Some had green seeds, and others had yellow seeds. Mendel wondered what caused the differences he observed and decided to conduct a series of simple experiments. He chose seven characteristics of pea plants for his experiments, as you can see in Table 3-1:

Table 3-1   Seven Traits of Pea Plants Studied by Gregor Mendel

Trait	Common Form	Uncommon Form
Seed color	Yellow	Green
Seed shape	Round	Wrinkled
Seed coat color	Gray	White
Pod color	Green	Yellow
Pod shape	Inflated	Constricted
Plant height	Tall	Short
Flower position	Along the stem	At the tip of the stem

For ten years, Mendel patiently grew many varieties of peas with various flower colors, seed shapes, seed numbers, and so on. In a process called *crossing,* he mated parent plants to see what their offspring would look like. When he passed away in 1884, Mendel was unaware of the magnitude of his contribution to science. A full 34 years passed after publication of his work (in 1868) before anyone realized what this simple gardener had discovered. (For the full story on how Mendel's research was lost and found again, flip to Chapter 22.)

If you don't know much about plants, understanding how plants reproduce may help you appreciate what Mendel did. To mate plants, you need flowers and the dusty substance they produce called *pollen* (the plant equivalent of sperm). Flowers have structures called *ovaries* (see Figure 3-1); the ovaries are hidden inside the *pistil* and are connected to the outside world by the *stigma*. Pollen is produced by structures called *stamen*. Like those of animals, the ovaries of plants produce eggs that, when exposed to pollen (in a process called *pollination*), are fertilized to produce seeds. Under the right conditions, the seeds sprout to become plants in their own right. The plants growing from seeds are the offspring of the plant(s) that produced the eggs and the pollen. Fertilization can happen in one of two ways:

- ✔ **Out-crossing:** Two plants are crossed, and the pollen from one can be used to fertilize the eggs of another.

- ✔ **Self-pollination (or selfing):** Some flowers produce both flowers and pollen, in which case the flower may fertilize its own eggs. Not all plants can self-fertilize, but Mendel's peas could.

Stamen   Stigma

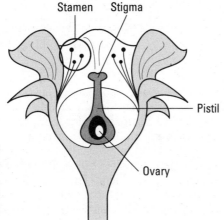

Pistil

Ovary

Figure 3-1:
Reproduc-
tive parts of
a flower.

Speaking the Language of Inheritance

You probably already know that genes are passed from parent to offspring and that somehow, genes are responsible for the physical traits (*phenotype,* such as hair color) you observe in yourself and the people and organisms around you. (For more on how genes do their jobs, you can flip ahead to Chapter 11.) The simplest possible definition of a *gene* is an inherited factor that determines some trait.

Genes come in different forms, called *alleles.* An individual's alleles determine the phenotype. The combinations of alleles of all the various genes that you possess make up your *genotype.* Genes occupy *loci* — specific locations along the strands of your DNA (*locus* is the singular form). Different traits (like hair texture and hair color) are determined by genes that occupy different loci, often on different chromosomes (see Chapter 2 for the basics of chromo-somes). Take a look at Figure 3-2 to see how alleles are arranged in various loci along two pairs of generic chromosomes.

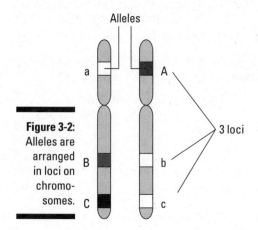

Figure 3-2:
Alleles are arranged in loci on chromo-somes.

In humans (and many other organisms), alleles of particular genes come in pairs. If both alleles are identical in form, that locus is said to be *homozygous,* and the whole organism can be called a *homozygote* for that particular locus. If the two alleles aren't identical, then the individual is *heterozygous,* or a *heterozygote,* for that locus. Individuals can be both heterozygous and homo-zygous at different loci at the same time, which is how all the phenotypic variation you see in a single organism is produced. For example, your hair texture is controlled by one locus, your hair color is controlled by different loci, and your skin color by yet other loci. You can see how figuring out how complex sets of traits are inherited would be pretty difficult.

Simplifying Inheritance

When it comes to sorting out inheritance, it's easiest to start out with how one trait is transmitted from one generation to the next. This is the kind of inheritance, sometimes called *simple inheritance,* that Mendel started with when first studying his pea plants.

Mendel's choice of pea plants and the traits he chose to focus on had posi-tive effects on his ability to uncover the laws of inheritance.

✔ **The original parent plants Mendel used in his experiments were true breeding.** When true breeders are allowed to self-fertilize, the exact same physical traits show up, unchanged, generation after generation.

True-breeding tall plants always produce tall plants, true-breeding short plants always produce short plants, and so on.

✔ **Mendel studied traits that had only two forms, or *phenotypes,* for each characteristic (like short or tall).** He deliberately chose traits that were either one type or another, like tall or short, or green-seeded or yellow-seeded. Studying traits that come in only two forms made the inheritance of traits much easier to sort out. (Chapter 4 covers traits that have more than two phenotypes.)

✔ **Mendel worked only on traits that showed an *autosomal dominant* form of inheritance — that is, the genes were located on autosomal (or non-sex) chromosomes.** (I discuss more complicated forms of inheritance in Chapters 4 and 5.)

Before his pea plants began producing pollen, Mendel opened the flower buds. He cut off either the pollen-producing part (the stamen) or the pollen-receiving part (the stigma) to prevent the plant from self-fertilizing. After the flower matured, he transferred pollen by hand — okay, not technically his hand; he used a tiny brush — from one plant (the "father") to another (the "mother"). Mendel then planted the seeds (the offspring) that resulted from this "mating" to see which physical traits each cross produced. The following sections explain the three laws of inheritance that Mendel discovered from his experiments.

Establishing dominance

For his experiments, Mendel crossed true-breeding plants that produced round seeds with true breeders that produced wrinkled seeds, crossed short true-breeders with tall true-breeders, and so on. Crosses of parent organisms that differ by only one trait, like seed shape or plant height, are called *monohybrid crosses.* Mendel patiently moved pollen from plant to plant, harvested and planted seeds, and observed the results after the offspring plants matured. His plants produced literally thousands of seeds, so his garden must have been quite a sight.

To describe Mendel's experiments and results, I refer to the parental generation with the letter *P.* I refer to the first offspring from a cross as *F1.* If F1 offspring are mated to each other (or allowed to self-fertilize), I call the next generation *F2* (see Figure 3-3 for the generation breakdown).

The results of Mendel's experiments were amazingly consistent. In every case when he mated true breeders of different phenotypes, all the F1 offspring had the same phenotype as one or the other parent plant. For example, when Mendel crossed a true-breeding tall parent with a true-breeding short parent, *all* the F1 offspring were tall. This result was surprising because until then, many people thought inheritance was a blending of the characteristics of the two parents — Mendel had expected his first generation offspring to be medium height.

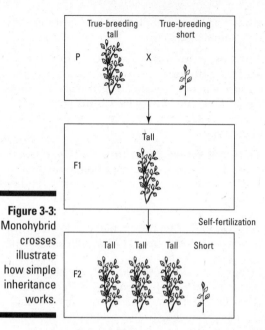

Figure 3-3:
Monohybrid
crosses
illustrate
how simple
inheritance
works.

If Mendel had just scratched his head and stopped there, he wouldn't have learned much. But he allowed the F1 offspring to self-fertilize, and something interesting happened: About 25 percent of the F2 offspring were short, and the rest, about 75 percent, were tall (refer to Figure 3-3).

From that F2 generation, when allowed to self-fertilize, his short plants were true breeders — all produced short progeny. His F2 tall plants produced both tall and short offspring. About one-third of his tall F2s bred true as tall. The rest produced tall and short offspring in a 3:1 ratio (that is, ¾ tall and ¼ short; refer to Figure 3-3).

After thousands of crosses, Mendel came to the accurate conclusion that the factors that determine seed shape, seed color, pod color, plant height, and so on are acting sets of two. He reached this understanding because *one* phenotype showed up in the F1 offspring, but *both* phenotypes were present among the F2 plants. The result in the F2 generation told him that whatever it was that controlled a particular trait (such as plant height) had been present but somehow hidden in the F1 offspring.

Mendel quickly figured out that certain traits seem to act like rulers, or dominate, other traits. *Dominance* means that one factor masks the presence of another. Round seed shape dominated wrinkled. Tall height dominated short. Yellow seed color dominated green. Mendel rightly determined the genetic principle of *dominance* by strictly observing phenotype in generation after generation and cross after cross. When true tall and short plants were crossed, each F1 offspring got one height-determining factor from each parent.

Because tall is *dominant* over short, all the F1 plants were tall. Mendel found that the only time *recessive* characters (traits that are masked by dominant traits) were expressed was when the two factors were alike, as when short plants self-fertilized.

Segregating alleles

Segregation is when things get separated from each other. In the genetic sense, what's separated are the two factors — the alleles of the gene — that determine phenotype. Figure 3-4 traces the segregation of the alleles for seed color through three generations. The shorthand for describing alleles is typically a capital letter for the dominant trait and the same letter in lowercase for the recessive trait. In this example, I use *Y* for the dominant allele that makes yellow seeds; *y* stands for the recessive allele that, when homozygous, makes seeds green.

The letters or symbols you use for various alleles and traits are arbitrary. Just make sure you're consistent in how you use letters and symbols, and don't get them mixed up.

In the segregation example featured in Figure 3-4, the parents (in the P generation) are homozygous. Each individual parent plant has a certain genotype — a combination of alleles — that determines its phenotype. Because pea plants are *diploid* (meaning they have two copies of each gene; see Chapter 2), the genotype of each plant is described using two letters. For example, a true-breeding yellow-seeded plant would have the genotype YY, and green-seeded plants are yy. The *gametes* (sex cells, as in pollen or eggs) produced by each plant bear only one allele. (Sex cells are *haploid;* see Chapter 2 for all the details on how meiosis produces haploid gametes.) Therefore, the true breeders can produce gametes of only one type — YY plants can only make Y gametes and yy plants can only produce y gametes. When a Y pollen and a y egg (or visa versa, y pollen and Y egg) get together, they make a Yy offspring — this is the heterozygous F1 generation.

The bottom line of the principle of segregation is this parsing out of the pairs of alleles into gametes. Each gamete gets one and only one allele for each locus; this is the result of homologous chromosomes parting company during the first round of meiosis (see Chapter 2 for more on how chromosomes split during meiosis). When the F1 generation self-fertilizes (to create the F2 generation), each plant produces two kinds of gametes: Half are Y, and the other half are y. Segregation makes four combinations of zygotes possible: YY, Yy, yY, or yy. (Yy and yY look redundant, but they're genetically significant because they represent different contributions [y or Y] from each parent.) Phenotypically, Yy, yY, and YY all look alike: yellow seeds. Only yy makes green seeds. The ratio of genotypes is 1:2:1 (¼ homozygous dominant: ½ heterozygous: ¼ homozygous recessive), and the ratio of phenotypes is 3 to 1 (dominant phenotype to recessive phenotype).

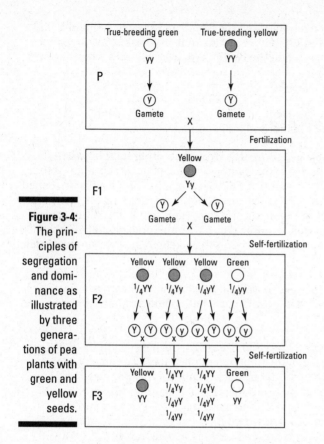

Figure 3-4:
The principles of segregation and dominance as illustrated by three generations of pea plants with green and yellow seeds.

If allowed to self-fertilize in the F3 generation, yy parents make yy offspring, and YY parents produce only YY offspring. The Yy parents again make YY, Yy, and yy offspring in the same ratios observed in the F2: ¼ YY, ½ Yy, and ¼ yy.

Scientists now know that what Mendel saw acting in sets of two were genes. Single pairs of genes (that is, one locus) control each trait. That means that plant height is at one locus, seed color at a different locus, seed shape at a third locus, and so on.

Declaring independence

As Mendel learned more about how traits were passed from one generation to the next, he carried out experiments with plants that differed in two or more traits. He discovered that the traits behaved independently — that is, that the inheritance of plant height had no effect on the inheritance of seed color, for example.

REMEMBER

The independent inheritance of traits is called the *law of independent assort-ment* and is a consequence of meiosis. When homologous pairs of chromo-somes separate, they do so randomly with respect to each other. The movement of each individual chromosome is independent with respect to every other chromosome. It's just like flipping a coin: As long as the coin isn't rigged, one coin flip has no effect on another — each flip is an independent event. Genetically, what this random separation amounts to is that alleles on different chromosomes are inherited independently.

Segregation and independent assortment are closely related principles. *Segregation* tells you that alleles at the same locus on pairs of chromosomes separate and that each offspring has the same chance of inheriting a particu-lar allele from a parent. *Independent assortment* means that every offspring also has the same opportunity to inherit any allele at any other locus (but this rule has some exceptions; see Chapter 4).

Finding Unknown Alleles

Mendel crossed parent plants in many different combinations to work out the identity of the hidden factors (which we now know as genes) that produced the phenotypes he observed. One type of cross was especially informative. A *testcross* is when any individual with an unknown genotype is crossed with a true-breeding individual with the recessive phenotype (in other words, a homozygote).

Each cross provides different information about the genotypes of the individ-uals involved. For example, Mendel could take any plant with any phenotype and testcross it with a true-breeding recessive plant to find out which alleles the plant of unknown genotype carried. Here's how the testcross would work: A plant with the dominant phenotype, violet flowers, could be crossed with a true-breeding white flowered plant (ww). If the resulting offspring all had violet flowers, Mendel knew that the unknown genotype was homozygous dominant (WW). In Figure 3-5, you see the results of another testcross: A heterozygote (Ww) testcross yielded offspring of half white and half violet phenotypes.

Figure 3-5:
The results of test-crosses divulge unknown genotypes.

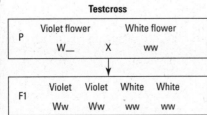

Applying Basic Probability to the Likelihood of Inheritance

Predicting the results of crosses is easy, because the rules of probability govern the likelihood of getting particular outcomes. The following are two important rules of probability that you should know:

- **Multiplication rule:** Used when the probabilities of events are independent of each other — that is, the result of one event doesn't influence the result of another. The combined probability of both events occurring is the product of the events, so you multiply the probabilities.

- **Addition rule:** Used when you want to know the probability of one event occurring as opposed to another, independent, event. Put another way, you use this rule when you want to know the probability of one *or* another event happening, but not necessarily both.

For more details about the laws of probability, check out the sidebar "Beating the odds with genetics."

Here's how you apply the addition and multiplication rules for monohybrid crosses (crosses of parent organisms that differ only by one trait). Suppose you have two pea plants. Both plants have violet flowers, and both are heterozygous (Ww). Each plant will produce two sorts of gametes, W and w, with equal probability — that is, half of the gametes will be W and half will be w for each plant. To determine the probability of a certain genotype resulting from the cross of these two plants, you use the multiplication rule and multiply probabilities. For example, what's the probability of getting a heterozygote (Ww) from this cross?

Because both plants are heterozygous (Ww), the probability of getting a W from plant one is $\frac{1}{2}$, and the probability of getting a w from plant two is also $\frac{1}{2}$. The word *and* tells you that you need to multiply the two probabilities to determine the probability of the two events happening together. So, $\frac{1}{2} \times \frac{1}{2} = \frac{1}{4}$. But there's another way to get a heterozygote from this cross: Plant one could contribute the w, and plant two could contribute the W. The probability of this turn of events is exactly equal to the first scenario: $\frac{1}{2} \times \frac{1}{2} = \frac{1}{4}$. Thus, you have two equally probable ways of getting a heterozygote: wW or Ww. The word *or* tells you that you must add the two probabilities together to get the total probability of getting a heterozygote: $\frac{1}{4} + \frac{1}{4} = \frac{1}{2}$. Put another way, there's a 50 percent probability of getting heterozygote offspring when two heterozygotes are crossed.

Beating the odds with genetics

When you try to predict the outcome of a certain event, like a coin flip or the gender of an unborn child, you're using probability. For many events, the probability is either-or. For instance, a baby can be either male or female, and a coin can land either heads or tails. Both outcomes are considered equally likely (as long as the coin isn't rigged somehow). For many events, however, determining the likelihood of a certain outcome is more complicated. Deciding how to calculate the odds depends on what you want to know.

Take, for example, predicting the sex of several children born to a given couple. The probability of any baby being a boy is ½, or 50 percent. If the first baby is a boy, the probability of the second child being a boy is still 50 percent, because the events that determine sex are independent from one child to the next (see Chapter 2 for a rundown of how meiosis works to produce gametes for sex cells). That means the sex of one child has no effect on the sex of the next child. But if you want to know the probability of having two boys in a row, you multiply the probability

of each independent event together: ½ × ½ = ¼, or 25 percent. If you want to know the probability of having two boys *or* two girls, you add the probabilities of the events together: ¼ (the probability of having two boys) + ¼ (the probability of having two girls) = ½, or 50 percent.

Genetic counselors use probability to determine the likelihood that someone has inherited a given trait and the likelihood that a person will pass on a trait if he or she has it. For example, a man and woman are each carriers for a recessive disorder, such as cystic fibrosis. The counselor can predict the likelihood that the couple will have an affected child. Just as in Mendel's flower crosses, each parent can produce two kinds of gametes, affected or unaffected. The man produces half-affected and half-unaffected gametes, as does the woman. The probability that any child inherits an affected allele from the mom *and* an affected allele from the dad is ¼ (that's ½ × ½). The probability that a child will be affected *and* female is ⅛ (that's ¼ × ½). The probability that a child will be affected *or* a boy is ¾ (that's ¼ + ½).

Solving Simple Genetics Problems

Every genetics problem, from those on an exam to one that determines what coat color your dog's puppies may have, can be solved in the same manner. Here's a simple approach to any genetics problem:

1. **Determine how many traits you're dealing with.**

2. **Count the number of phenotypes.**

3. **Carefully read the problem to identify the question.** Do you need to calculate genetic or phenotypic ratios? Are you trying to determine something about the parents or the offspring?

4. **Look for words that mean *and* and *or* to help determine which probabilities to add and which to multiply.**

Deciphering a monohybrid cross

Imagine that you have your own garden full of the same variety of peas that Mendel studied. After reading this book, filled with enthusiasm for genetics, you rush out to examine your pea plants, having noticed that some plants are tall and others short. You know that last year you had one tall plant (which self-fertilized) and that this year's crop consists of the offspring of last year's one tall parent plant. After counting plants, you discover that 77 of your plants are tall, and 26 are short. What was the genotype of your original plant? What is the dominant allele?

You have two distinct phenotypes (tall and short) of one trait — plant height. You can choose any symbol or letter you please, but often, geneticists use a letter like *t* for short and then capitalize that letter for the other allele (here, *T* for tall).

One way to start solving the problem of short versus tall plants is to determine the ratio of one phenotype to the other. To calculate the ratios, add the number of offspring together: 77 + 26 = 103, and divide to determine the proportion of each phenotype: 77 ÷ 103 = 0.75, or 75 percent are tall. To verify your result, you can divide 26 by 103 to see that 25 percent of the offspring are short, and 75 percent plus 25 percent gives you 100 percent of your plants.

From this information alone, you've probably already realized (thanks to simple probability) that your original plant must have been heterozygous and that tall is dominant over short. As I explain in the "Segregating alleles" section earlier in this chapter, a heterozygous plant (Tt) produces two kinds of gametes (T or t) with equal probability (that is, half the time the gametes are T and the other half they're t). The probability of getting a homozygous dominant (TT) genotype is $\frac{1}{2} \times \frac{1}{2} = \frac{1}{4}$ (that's the probability of getting T twice: T once *and* T a second time, like two coin flips in a row landing heads). The probability of getting a heterozygous dominant (T and t, *or* t and T) is $\frac{1}{2} \times \frac{1}{2} = \frac{1}{4}$ (to get Tt) plus $\frac{1}{2} \times \frac{1}{2} = \frac{1}{4}$ (tT). The total probability of a plant with the dominant phenotype (TT *or* Tt *or* tT) is $\frac{1}{4} + \frac{1}{4} + \frac{1}{4} = \frac{3}{4}$. With 103 plants, you'd expect 77.25 (on average) of them to show the dominant phenotype — which is essentially what you observed.

Tackling a dihybrid cross

To become more comfortable with the process of solving simple genetics problems, you can tackle a problem that involves more than one trait: a *dihybrid cross*.

Here's the problem scenario. In bunnies, short hair is dominant. (If you're a rabbit breeder, please forgive my oversimplification.) Your roommate moves out and leaves behind two bunnies (you were feeding them anyway, and they're cute, so you don't mind). One morning you wake to find that your bunnies are now parents to a litter of babies.

> ✔ One is gray and has long fur.
>
> ✔ Two are black and have long fur.
>
> ✔ Two are gray and have short fur.
>
> ✔ Seven look just like the parents: black with short fur.

Besides the meaningful lesson about spaying and neutering pets, what can you discover about the genetics of coat color and hair length of your rabbits?

First, how many traits are you dealing with? I haven't told you anything about the gender of your baby bunnies, so it's safe to assume that sex doesn't have anything to do with the problem. (I take that back. Sex is the source of the problem — see Chapter 5 for more on the genetics of sex.) You're dealing with two traits: color of fur and length of fur. Each trait has two phenotypes: Fur can be black or gray, and length of fur can be long or short. In working through this problem, you're told upfront that short fur is a dominant trait, but you don't get any information about color.

The simplest method is to examine one trait at a time — in other words, look at the monohybrid crosses. (Jump back to the section "Deciphering a monohybrid cross" for a refresher.)

Both parents have short fur. How many of their offspring have short fur? Nine of twelve, and 9 ÷ 12 = ¾, or 75 percent. That means there are three shorthaired bunnies to every one long-haired bunny.

Being identical in phenotype, the parents both have black coats. How many babies have black coats? Nine of twelve. There's that comfortingly familiar ratio again! The ratio of black to gray is 3 to 1.

From your knowledge of monohybrid crosses, you've probably guessed that the parent rabbits are heterozygous for coat color and, at the same time, are heterozygous for fur length. To be sure, you can calculate the probability of certain genotypes and corresponding phenotypes of offspring for two rabbits that are heterozygous at two loci (see Figure 3-6).

The phenotypic ratio observed in the rabbits' offspring (9:3:3:1; refer to Figure 3-6) is typical for the F2 generation in a dihybrid cross. The rarest phenotype is the one that's recessive for both traits; in this case, long hair and gray color are both recessive. The most common phenotype is the one that's dominant for both traits. The fact that seven of your twelve baby rabbits are black with short fur tells you that the probability of getting a particular allele for color and a particular allele for coat length is the product of two independent events. Coat color and hair length are coded by genes that are inherited independently — as you would expect under the principle of independent assortment.

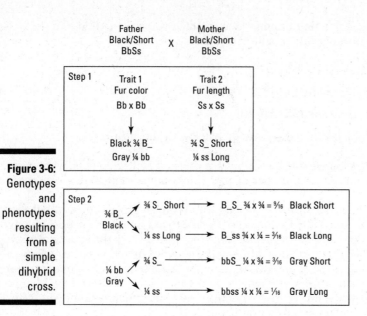

Figure 3-6:
Genotypes
and
phenotypes
resulting
from a
simple
dihybrid
cross.

Chapter 4

Law Enforcement: Mendel's Laws Applied to Complex Traits

Although nearly 150 years have elapsed since Gregor Mendel cultivated his pea plants (see Chapter 3), the observations he made and the conclusions he drew still accurately describe how genes are passed from parent to offspring. The basic laws of inheritance — dominance, segregation, and independent assortment — continue to stand the test of time.

However, inheritance isn't nearly as simple as Mendel's experiments suggest. Dominant alleles don't always dominate, and genes aren't always inherited independently. Some genes mask their appearances, and some alleles can kill. This chapter explains exactly how Mendel was right, and wrong, about the laws of inheritance and how they're enforced.

Dominant Alleles Rule . . . Sometimes

If Mendel had chosen a plant other than the pea plant for his experiments, he may have come to some very different conclusions. The traits that Mendel studied show *simple dominance* — when the dominant allele's *phenotype,* or physical trait (a yellow seed, for example), masks the presence of the recessive allele. The recessive phenotype (a green seed in this example) is only expressed when both alleles are recessive, which is written as *yy.* (Turn to Chapter 3 for the definitions of commonly used genetics terms such as *allele, recessive,* and *homozygote.*) But not all alleles behave neatly as dominant-recessive. Some alleles show incomplete dominance and therefore seem to display a blend of phenotypes from the parents. This section tells you how dominant alleles rule the roost — but only part of the time.

Wimping out with incomplete dominance

A trip to the grocery store can be a nice genetics lesson. Take eggplant, for example. Eggplant comes in various shades of (mostly) purple skin that are courtesy of a pair of alleles at a single locus interacting in different ways to express the phenotype — purple fruit color. Dark purple and white colors are both the result of homozygous alleles. Dark purple is homozygous for the dominant allele (PP), and white is homozygous for the recessive allele (pp). When crossed, dark purple and white eggplants yield light purple offspring — the intermediate phenotype. This intermediate color is the result of the allele for purple being incomplete in its dominance of the allele for white (which is actually the allele for no color).

With *incomplete dominance,* the alleles are inherited in exactly the same way they always are: One allele comes from each parent. The alleles still conform to the principles of segregation and independent assortment, but the way those alleles are expressed (the phenotype) is different. (You can find out about exceptions to the independent assortment rules in the section "Genes linked together" later in this chapter.)

Here's how the eggplant cross works: The parent plants are PP (for purple) and pp (for white). The F1 generation is all heterozygous (Pp), just as you'd expect from Mendel's experiments (see Chapter 3). If this were a case of simple dominance, all the Pp F1 generation would be dark purple. But in this case of incomplete dominance, the F1 generation comes out light purple (sometimes called violet). (The heterozygotes produce a less purple pigment, making the offspring lighter in color than homozygous purple plants.)

In the F2 (the result of crossing Pp with Pp), half the offspring have violet fruits (corresponding with the Pp genotype). One-quarter of the offspring are dark purple (PP) and one-quarter are white (pp) — these are the homozygous offspring. Rather than the 3:1 phenotypic ratio (three dark purple eggplants and one white eggplant) you'd expect to see with simple dominance, with incomplete dominance, you see a 1:2:1 ratio (one dark purple eggplant, two light purple eggplants, and one white eggplant) — the exact ratio of the underlying genotype (PP, Pp, Pp, pp).

Keeping it fair with codominance

When alleles share equally in the expression of their phenotypes, the inheritance pattern is considered *codominant.* Both alleles are expressed fully as phenotypes instead of experiencing some intermediate expression (like what's observed in incomplete dominance).

You can see a good example of codominance in human blood types. If you've ever donated blood (or received a transfusion), you know that your blood type is extremely important. If you receive the wrong blood type during a transfusion, you can have a fatal allergic reaction. Blood types are the result of proteins, called *antigens,* that your body produces on the surface of red blood cells. Antigens protect you from disease by recognizing invading cells (like bacteria) as foreign, and then binding to the cells and destroying them.

Your antigens determine your blood type. Several alleles code for blood antigens. Dominant alleles code two familiar blood types, A and B. When a person has both A and B alleles, the person's blood produces both antigens simultaneously and in equal amounts. Therefore, a person who has an AB genotype also has the AB phenotype.

The situation with ABO blood types gets more complicated by the presence of a third allele for type O in some people. The O allele is recessive, so ABO blood types show two sorts of inheritance:

✔ Codominance (for A and B)

✔ Dominant-recessive (A or B paired with the O allele)

Type O is only expressed in the homozygous state. For more information on multiple alleles, check out the section "More than two alleles" later in this chapter.

Dawdling with incomplete penetrance

Some dominant alleles don't express their influence consistently. When dominant alleles are present but fail to show up as a phenotype, the condition is termed *incompletely penetrant. Penetrance* is the probability that an individual having a dominant allele will show the associated phenotype. *Complete penetrance* means every person having the allele shows the phenotype. Most dominant alleles have 100 percent penetrance — that is, the phenotype is expressed in every individual possessing the allele. However, other alleles may show reduced, or incomplete, penetrance, meaning that individuals carrying the allele have a reduced probability of having the trait.

Penetrance of disease-causing alleles like those responsible for certain cancers or other hereditary disorders complicate matters in genetic testing (see Chapter 12 to find out more about genetic testing for disease). For example, one of the genes associated with breast cancer *(BRCA1)* is incompletely penetrant. Studies estimate that approximately 70 percent of women carrying the allele will be affected by breast cancer by age 70. Therefore, genetic tests indicating that someone carries the allele only point to increased risk, not a certainty of getting the disease, and indicate a need for affected women to be screened regularly for early signs of the disease, when treatment can be most effective.

Geneticists usually talk about penetrance in terms of a percentage. In this example, the breast cancer gene is 70 percent penetrant.

Regardless of penetrance, the degree to which an allele expresses the phenotype may differ from individual to individual; this variable strength of a trait is called *expressivity*. One trait with variable expressivity that shows up in humans is *polydactyly,* the condition of having more than ten fingers or toes. In persons with polydactyly, the expressivity of the trait is measured by the completeness of the extra digits — some people have tiny skin tags, and others have fully functional extra fingers or toes.

Alleles Causing Complications

The variety of forms that genes (as alleles) take accounts for the enormous diversity of physical traits you see in the world around you. For example, many alleles exist for eye color and hair color. In addition, several loci contribute to most phenotypes. Dealing with multiple loci and many alleles at each locus complicates inheritance patterns and makes them harder to understand. For many disorders, scientists don't fully understand the form of inheritance because variable expressivity and incomplete penetrance mask the patterns. Additionally, multiple alleles can interact as incompletely dominant, codominant, or dominant-recessive (see "Dominant Alleles Rule . . . Sometimes" earlier in this chapter for the whole story). This section explains how various alleles of a single gene can complicate inheritance patterns.

More than two alleles

When it came to his pea plant research, Mendel deliberately chose to study traits that came in only two flavors. For instance, his peas had only two flower color possibilities: white and purple. The allele for purple in the common pea plant is fully dominant, so it shows up as the same shade of purple in both heterozygous and homozygous plants. In addition to being fully dominant, purple is completely penetrant, so every single plant that inherits the gene for purple flowers has purple flowers.

If Mendel had been a rabbit breeder instead of a gardener, his would likely be a different story. He may not have earned the title "Father of Genetics," because the broad spectrum of rabbit coat colors would make most anyone simply throw up his hands.

To simplify matters, consider one gene for coat color in bunnies. The C gene has four alleles that control the amount of pigment produced in the hair shaft. These four alleles give you four rabbit color patterns to work with. The various rabbit color alleles are designated by the letter c with superscripts:

✔ **Brown (c⁺):** Brown rabbits are considered *wild-type,* which is generally considered the "normal" phenotype. Brown rabbits are brown all over.

✔ **Albino (c):** Rabbits homozygous for this color allele don't produce any pigment at all. Therefore, these white rabbits are considered *albino.* They have all-white coats, pink eyes, and pink skin.

✔ **Chinchilla (c^{ch}):** Chinchilla rabbits are solid gray (specifically, they have white hair with black tips).

✔ **Himalayan (c^h):** Himalayan rabbits are white but have dark hair on their feet, ears, and noses.

Wild-type is a bit of a problematic term in genetics. Generally, wild-type is considered the "normal" phenotype, and everything else is "mutant." *Mutant* is simply different, an alternative form that's not necessarily harmful. Wild-type tends to be the most common phenotype and is usually dominant over other alleles. You're bound to see wild-type used in genetics books to describe phenotypes such as eye color in fruit flies. Though rare, the mutant color forms occur in natural populations of animals. In the case of domestic rabbits, color forms other than brown are the product of breeding programs specifically designed to obtain certain coat colors.

Although a particular trait can be determined by a number of different alleles (as in the four allele possibilities for rabbit coat color), any particular animal carries only two alleles at a particular locus at one time.

The C gene in rabbits exhibits a dominance hierarchy common among genes with multiple alleles. Wild-type is completely dominant over the other three alleles, so any rabbit having the c⁺ allele will be brown. Chinchilla is incompletely dominant over Himalayan and albino. That means heterozygous chinchilla/Himalayan rabbits are gray with dark ears, noses, and tails. Heterozygous chinchilla/albinos are lighter than homozygous chinchillas. Albino is only expressed in animals that are homozygous (cc).

The color alleles in monohybrid crosses for rabbit color follow the same rules of segregation and independent assortment that apply to the pea plants that Mendel studied (see Chapter 3). The phenotypes for rabbit color are just more complex. For example, if you were to cross an albino rabbit (cc) with a homozygous chinchilla (c^{ch}c^{ch}), in the F2 generation (cc^{ch} mated with cc^{ch}) you'd get the expected 1:2:1 genotypic ratio (1 cc to 2 cc^{ch} to 1 c^{ch}c^{ch}); the phenotypes would show a corresponding 1:2:1 ratio (one albino, two light chinchilla, one full chinchilla).

A total of five genes actually control coat color in rabbits. The section "Genes in hiding" later in this chapter delves into how multiple genes interact to create fur color.

Lethal alleles

Many alleles express unwanted traits (phenotypes) that indirectly cause suffering and death (such as the excessive production of mucus in the lungs of cystic fibrosis patients). Rarely, alleles may express the *lethal phenotype* — that is, death — immediately and thus are never expressed beyond the zygote. These alleles produce a 1:2 phenotypic ratio, because only heterozygotes and homozygous nonlethals survive to be counted.

The first lethal allele that scientists described was associated with yellow coat color in mice. Yellow mice are *always* heterozygous. When yellow mice are bred to other yellow mice, they produce yellow and non-yellow offspring in a 2:1 ratio, because all homozygous yellow mice die as embryos. Homozygous yellow has no real phenotype (beyond dead), because these animals never survive.

Lethal alleles are almost always recessive, and thus are expressed only in homozygotes. One notable exception is the gene that causes Huntington disease. Huntington disease (also known as Huntington chorea) is inherited as an autosomal dominant lethal disorder, meaning that persons with Huntington disease develop a progressive nerve disorder that causes involuntary muscle movement and loss of mental function. Huntington disease is expressed in adulthood and is always fatal. It has no cure; treatment is aimed at alleviating symptoms of the disease.

Making Life More Complicated

Many phenotypes are determined by the action of more than one gene at a time. Genes can hide the effects of each other, and sometimes one gene can control several phenotypes at once. This section looks at how genes make life more complicated (and more interesting).

When genes interact

If you don't mind returning to the produce section of your local grocery store (no more eggplants, I promise), you can observe the interaction of multiple genes to produce various colors of bell peppers. Two genes (R and C) interact to make these mild, sweet peppers appear red, brown, yellow, or green. You see four phenotypes as the result of two alleles at each locus.

Figure 4-1 shows the genetic breakdown of bell peppers. In the parental generation (P), you start with a homozygous dominant pepper (RRCC), which is red, crossed with a homozygous recessive (rrcc) green pepper. (This is a dihybrid cross — that is, one involving two genes — like the one I describe at the end of Chapter 3.) You can easily determine the expected genotypic ratios by considering each locus separately. For the F1 generation, that's really easy to do, because both loci are heterozygous (RrCc). Just like homozygous dominant peppers, fully heterozygous peppers are red. When the F1 peppers self-fertilize, the phenotypes of brown and yellow show up.

Brown pepper color is produced by the genotype R_cc. The blank means that the R locus must have at least one dominant allele present to produce color, but the other allele can be either dominant or recessive. Yellow is produced by the combination rrC_. To make yellow pigment, the C allele must be either heterozygous dominant or homozygous dominant with a recessive homozygous R allele. The F2 generation shows the familiar 9:3:3:1 dihybrid phenotypic ratio (just like the guinea pigs do in Chapter 3). The loci assort independently, just as you'd expect them to.

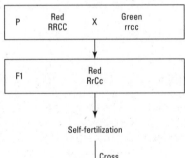

Figure 4-1: Genes interact to produce pigment in this dihybrid cross for pepper color.

Conclusion: $\frac{9}{16}$ Red $\frac{3}{16}$ Brown $\frac{3}{16}$ Yellow $\frac{1}{16}$ Green

Genes in hiding

As the preceding section explains, in pepper color, the alleles of two genes interact to produce color. But sometimes, genes hide or mask the action of other genes altogether. This occurrence is called *epistasis.*

A good example of epistasis is the way in which color is determined in horses. Like that of dogs, cats, rabbits, and humans, hair color in horses is determined by numerous genes. At least seven loci determine color in horses. To simplify mastering epistasis, you tackle the actions of only three genes: W, E, and A (see Table 4-1 for a rundown of the genes and their effects). One locus (W) determines the presence or absence of color. Two loci (E and A) interact to determine the distribution of red and black hair — the most common hair colors in horses.

A horse that carries one dominant allele for W will be albino — no color pigments are produced, and the animal has white skin, white hair, and pink eyes. (Homozygous dominant for the white allele is lethal; therefore, no living horse possesses the WW genotype.) All horses that are some color other than white are homozygous recessive (ww). (If you're a horse breeder, you know that I'm really oversimplifying here. Please forgive me.) Therefore, the dominant allele W shows *dominant epistasis* because it masks the presence of other alleles that determine color.

If a horse isn't white (that is, not albino), then two main genes are likely determining its hair color: E and A. When the dominant allele E is present, the horse has black hair (it may not be black all over, but it's black somewhere). Black hair is expressed because the E locus controls the production of two pigments, red and black. EE and Ee horses produce both black and red pigments. Homozygous recessive (ee) horses are red; in fact, they're always red regardless of what's happening at the A locus. Thus, ee is *recessive epistatic,* which means that in the homozygous recessive individual, the locus masks the action of other loci. In this case, the production of black pigment is completely blocked.

When a horse has at least one dominant allele at the E locus, the A locus controls the amount of black produced. The A locus (also called *agouti,* which is a dark brown color) controls the production of black pigments. A horse with the dominant A allele produces black only on certain parts of its body (often on its mane, tail, and legs — a pattern referred to as *bay*). Horses that are aa are simply black. However, the homozygous recessive E locus (*ee*) masks the A locus entirely (regardless of genotype), blocking black color completely.

Table 4-1	Genetics of Hair Color in Horses		
Genotype	*Phenotype*	*Type of Epistasis*	*Effect*
WW__	Lethal	No epistasis	Death
Ww__	Albino	Dominant	Blocks all pigments
wwE_aa	Black	Recessive	Blocks red
wwE_A_	Bay or brown	No epistasis	Both red and black expressed
wwee__	Red	Recessive	Blocks black

This example of the genetics of horse hair color proves that the actions of genes can be complex. In this one example, you see a lethal allele (W) along with two other loci that can each mask the other under the right combination of alleles. This potential explains why it can be so difficult to determine how certain conditions are inherited. Epistasis can act along with reduced penetrance to create extremely elusive patterns of inheritance — patterns that often can only be worked out by examining the DNA itself. (I cover genetic testing in Chapter 12.)

Genes linked together

Roughly 30 years after Mendel's work was rediscovered in 1900 and verified by the scientific community (see Chapter 22 for the whole story), the British geneticist Ronald A. Fisher realized that Mendel had been exceptionally lucky — either that or he'd cheated. Of the many, many traits Mendel could have studied, he published his results on seven traits that conform to the laws of segregation and independent assortment, have two alleles, and show dominant-recessive inheritance patterns. Fisher asserted that Mendel must have published the part of his data he understood and left out the rest. (After Mendel died, all his papers were burned, so we'll never know the truth.) The rest would include all the parts that make inheritance messy, like epistasis and *linkage*.

Because of the way genes are situated along chromosomes, genes that are very close together spatially (that is, fewer than 50 million base pairs apart; see Chapter 6 for how DNA is measured in base pairs) are inherited together. When genes are so close together that they're inherited together (either all or part of the time), the genes are said to be *linked* (see Figure 4-2). The occurrence of linked genes means that not all genes are subject to independent assortment. To determine if genes are linked, geneticists carry out a process called *linkage analysis*.

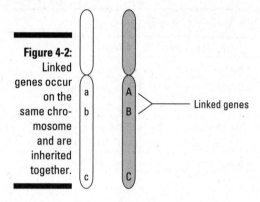

Figure 4-2:
Linked
genes occur
on the
same chro-
mosome
and are
inherited
together.

Linked genes

The process of linkage analysis is really a determination of how often *recombination* (the mixing of information, also called *crossing-over,* contained on the two homologous chromosomes; see Chapter 2) occurs between two or more genes. If genes are close enough together on the chromosome, they end up being linked more than 50 percent of the time. However, genes on the same chromosome can behave as if they were on different chromosomes, because during the first stage of meiosis (see Chapter 2), crossing-over occurs at many points along the two homologous chromosomes. If crossing-over splits two loci up more than 50 percent of the time, the genes on the same chromosome appear to assort independently, as if the genes were on different chromosomes altogether.

Generally, geneticists perform linkage analysis by examining dihybrid crosses (dihybrid means two loci; see Chapter 3) between a heterozygote and a homozygote. If you want to determine the linkage between two traits in fruit flies, for example, you choose an individual that's AaBb and cross it with one that's aabb. If the two loci, A and B, are assorting independently, you can expect to see the results shown in Figure 4-3. The heterozygous parent produces four types of gametes — AB, aB, Ab, and ab — with equal frequency. The homozygous parent can only make one sort of gamete — ab. Thus, in the F1 offspring, you see a 1:1:1:1 ratio.

But what if you see a completely unexpected ratio, like the one shown in Table 4-2? What does that mean? These results indicate that the traits are linked.

As you can see in Figure 4-4, the dihybrid parent makes four sorts of gametes. Even though the loci are on the same chromosome, the gametes don't occur in equal frequency. Most of the gametes show up just as they do on the chromosome, but crossover occurs between the two loci roughly 20 percent of the time, producing the two rarer sorts of gametes (each is produced about 10 percent of the time). Crossover occurs with roughly the same frequency in the homozygous parent, too, but because the alleles are the same, the results of those crossover events are invisible. Therefore, you can safely ignore that part of the problem.

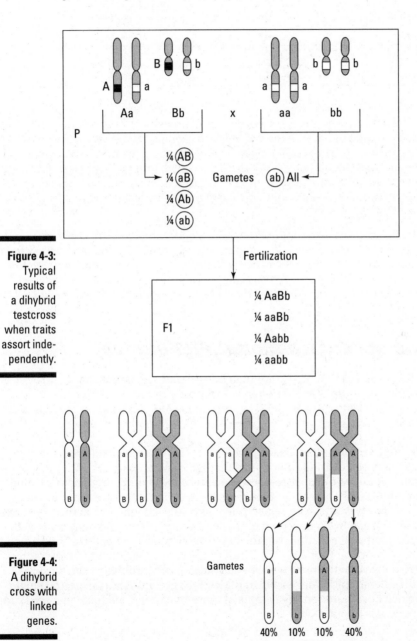

Figure 4-3:
Typical results of a dihybrid testcross when traits assort independently.

Figure 4-4:
A dihybrid cross with linked genes.

Table 4-2	Linked Traits in a Dihybrid Testcross	
Genotype	**Number of Offspring**	**Proportion**
Aabb	320	40%
aaBb	318	40%
AaBb	80	10%
aabb	76	10%

To calculate *map distance,* or the amount of crossover, between two loci, you divide the total number of recombinant offspring by the total number of offspring observed. The *recombinant offspring* are the ones that have a genotype different from the parental genotype. This calculation gives you a proportion: percent recombination. One map unit distance on a chromosome is equal to 1 percent recombination. Generally, one map unit is considered to be 1 million base pairs long.

As it turns out, genes for four of the traits Mendel studied were situated together on chromosomes. Two genes were on chromosome 1, and two were on chromosome 4; however, the genes were far enough apart that recombination was greater than 50 percent. Thus, all four traits appeared to assort independently, just as they would have if they'd been on four different chromosomes.

One gene with many phenotypes

Certain genes can control more than one phenotype. Genes that control multiple phenotypes are *pleiotropic.* Pleiotropy is very common; almost any major single gene disorder listed in the Online Mendelian Inheritance in Man database (www.ncbi.nlm.nih.gov/omim) shows pleiotropic effects.

Take, for example, phenylketonuria (PKU). This disease is inherited as a single gene defect and is autosomal recessive. When persons with the homozygous recessive phenotype consume substances containing phenylalanine, their bodies lack the proper biochemical pathway to break down the phenylalanine into tyrosine. As a result, phenylalanine accumulates in the body, preventing normal brain development. The primary phenotype of persons with PKU is mental retardation, but the impaired biochemical pathway affects other phenotypic traits as well. Thus, PKU patients also exhibit light hair color, unusual patterns of walking and sitting, skin problems, and seizures. All the phenotypic traits associated with PKU are associated with the single gene defect rather than the actions of more than one gene (see Chapter 12 for more details about PKU).

Uncovering More Exceptions to Mendel's Laws

As inheritance of genetic disorders is better studied, many exceptions to strict Mendelian inheritance rules arise. This section addresses four important exceptions.

Epigenetics

One of the biggest challenges to Mendel's laws comes from a phenomenon called *epigenetics.* The prefix *epi-* means "over" or "above." In epigenetics, organisms with identical alleles (including identical twins) may exhibit different phenotypes.

The difference in phenotypes doesn't come from the genes themselves but from elsewhere in the chemical structure of the DNA molecule (you can find out all about DNA's chemical and physical structure in Chapter 6). What happens is that tiny chemical tags, called *methyl groups,* are attached to the DNA. In essence, the tags act like the operating system in your computer that tells the programs how often to work, where, and when. In the case of epigenetics, the tags can shut genes down or turn genes on. Not only that, but the tags are inherited by the next generation as well.

Some epigenetic effects are normal and useful: They control how your various cells look and behave, like the differences between a heart muscle cell and a skin cell. However, other tags act like mutations and cause diseases like cancer (discover more about the role DNA plays in cancer in Chapter 14). Epigenetics is an exciting area of genetics research that will yield answers to how the genetic code in your DNA is affected by aging, your environment, and much more.

Genomic imprinting

Genomic imprinting is a special case of epigenetics. When traits are inherited on autosomal chromosomes, they're generally expressed equally in males and females. In some cases, the gender of the parent who contributes the particular allele may affect how the trait is expressed; this is called *genomic imprinting.*

Sheep breeders in Oklahoma discovered an amusing example of genomic imprinting. A ram named Solid Gold had unusually large hindquarters for his breed. Eventually, Solid Gold sired other sheep, which also had very large . . . butts. The breed was named Callipyge, which is Greek for *beautiful butt*. It turns out that six genes affect rump size in sheep. As breeders mated Callipyge sheep, it quickly became clear that the trait didn't obey Mendel's rules. Eventually, researchers determined that the big rump phenotype resulted only when the father passed on the trait. Callipyge ewes can't pass their big rumps on to their offspring.

The reasons behind genomic imprinting are still unclear. In the case of Callipyge sheep, scientists think there may be a mutation in a gene that regulates other genes, but why the expression of the gene is controlled by only paternal chromosomes remains a mystery. (Genomic imprinting is a big issue in cloning as well; see Chapter 20 for more on that topic.)

Anticipation

Sometimes, traits seem to grow stronger and gain more expressivity from one generation to the next. The strengthening of a trait as it's inherited is called *anticipation*. Schizophrenia is a disorder that's highly heritable and often shows a pattern of anticipation. It affects a person's mood and how she views herself and the world. Some patients experience vivid hallucinations and delusions that lead them to possess strongly held beliefs such as paranoia or grandeur. The age of onset of schizophrenic symptoms and the strength of the symptoms tend to increase from one generation to the next.

The reason behind anticipation in schizophrenia and other disorders, such as Huntington disease, may be that during replication (covered in Chapter 7), repeated sections of the DNA within the gene are easily duplicated by accident (see Chapter 13 for more on mutation by duplication). Thus, in successive generations, the gene actually gets longer. As the gene grows longer, its effects get stronger as well, leading to anticipation. In disorders affecting the brain, the mutation leads to malformed proteins (see Chapter 9 for how genes are translated into protein). The malformed proteins accumulate in the brain cells, eventually causing cells to die. Because the malformed proteins may get larger in successive generations, the effects show up when the person is young or they manifest themselves as a more severe form of the disease.

Environmental effects

Most traits show little evidence of environmental effect. However, the environment that some organisms live in controls the phenotype that some of its genes produce. For example, the gene that gives a Himalayan rabbit its characteristic phenotype of dark feet, ears, nose, and tail is a good example of a trait that varies in its expression based on the animal's environment. The pigment that produces dark fur in any animal results from the presence of an enzyme that the animal's body produces. But in this case, the enzyme's effect is deactivated at normal body temperature. Thus, the allele that produces pigment in the rabbit's fur is expressed only in the cooler parts of the body. That's why Himalayan rabbits are all white when they're born (they've been kept warm inside their mother's body) but get dark feet, ears, noses, and tails later in life. (Himalayans also change color seasonally and get lighter during the warmer months.)

Phenylketonuria (see "One gene with many phenotypes" earlier in this chapter) and other disorders of metabolism also depend on environmental factors — such as diet — for the expression of the trait.

Chapter 5

Differences Matter: The Genetics of Sex

- -

In This Chapter

▶ How sex is determined in humans and other animals

▶ What sorts of disorders are associated with sex chromosomes

▶ How sex affects other traits

- -

*S*ex is a term with many meanings. For geneticists, sex usually refers to two related concepts: the phenotype of sex (either male or female) and reproduction. It's hard to underestimate the importance of sex when it comes to genetics. Sex influences the inheritance of traits from one generation to the next and how those traits are expressed. Sexual reproduction allows organisms to create an amazing amount of genetic diversity via their offspring, which is handy because genetically diverse populations are more resilient in the face of disease and disaster. Many different individuals carrying many different alleles of the same genes increases the likelihood that some individuals will be resistant to disease and the effects of disaster and will pass that resistance on to their offspring. (For more on the importance of genetic diversity, flip to Chapter 17.)

In this chapter, you discover how chromosomes act to determine sex in humans and other organisms, how sex influences the expression of various nonsex (autosomal) traits, and what happens when too many or too few sex chromosomes are present.

X-rated: How You Got So Sexy

Presumably since the beginning of time, humans have been aware of the dissimilarities between the sexes. But it wasn't until 1905 that Nettie Stevens stared through a microscope long enough to discover the role of the Y chromosome in the grand scheme of things. Until Stevens came along, the much larger X chromosome was credited with creating all the celebrated differences between males and females.

From a genetics standpoint, the phenotypes of sex — male and female — depend on which type of gamete an individual produces. If an individual produces sperm (or has the potential to, when mature), it's considered male. If the individual can produce eggs, it's considered female. Some organisms are both male and female (that is, they're capable of producing viable eggs and sperm); this situation is referred to as *monoecy* (pronounced mo-*knee*-see, which means "one house"). Many plants, fish, and invertebrates (organisms lacking a bony spine like yours) are *monoecious* (mo-*knee*-shus).

Humans are *dioecious* (*di*-ee-shus; literally "two houses"), meaning that individuals have either functional male or female reproductive structures, but not both. Most of the species you're familiar with are dioecious: Mammals, insects, birds, reptiles, and many plants all have separate genders.

Organisms with separate genders get their sex phenotypes in various ways.

- ✔ Chromosomal sex determination occurs when the presence or absence of certain chromosomes control sex phenotype.
- ✔ Genetic sex determination occurs when particular genes control sex phenotype.
- ✔ The environment an organism develops in may determine its gender.

This section examines how chromosomes, genetics, and the environment determine whether an organism is male or female.

Sex determination in humans

In humans and most other mammals, males and females have the same number of chromosomes (humans have 46) in pairs (making humans *diploid*). Sex phenotype is determined by two sex chromosomes: X and Y. (Figure 5-1 shows the basic size and shape of these chromosomes.) Female humans have two X chromosomes, and male humans have one X and one Y. Check out the sidebar "The X (and Y) files" for how X and Y got their names. (Chromosomes have their stereotypical sausage shapes only during metaphase of mitosis or meiosis. Check out Chapter 2 for more details on mitosis and meiosis.)

The very important X

During metaphase, the X chromosome truly has an x-shape, with the centromere placed roughly in the middle (see Chapter 2 for more about chromosomes and their shapes). Genetically speaking, unlike the relatively puny Y chromosome, X is quite large. Of the 23 pairs of chromosomes ordered by size, X occupies the eighth place, weighing in at slightly over 150 million base pairs long. (See Chapter 6 for more about how DNA is measured in base pairs.)

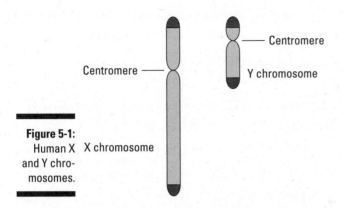

Centromere
Centromere
Y chromosome

Figure 5-1:
Human X
and Y chro-
mosomes.

X chromosome

The X chromosome is home to between 900 and 1,200 genes and is incredibly important for normal human development. When no X is present, the zygote can't commence development. Table 5-1 lists a few of X's genes that are required for survival. Surprisingly, only one gene on X has a role in determining female phenotype; all the other genes that act to make females are on the autosomal (nonsex) chromosomes.

Table 5-1	Important Genes on the X Chromosome
Gene	*Function*
ALAS2	Directs formation of red blood cells
ATP7A	Regulates copper levels in the body
COL4A5	Required for normal kidney function
DMD	Controls muscle function and pathways between nerve cells
F8	Responsible for normal blood clotting

In all mammals (including humans), the developing embryo starts in what developmental biologists refer to as an *indifferent stage,* meaning the embryo has the potential to be either male or female. Here's how sex determination in mammals works: In roughly the fourth week of development, the embryo begins to develop a region near the kidneys called the *genital ridge.* Three genes (all on autosomes) kick in to convert the genital ridge tissue into tissue that can become sex organs. The tissue that's present by week seven in the embryo's development is called the *bipotential gonad* because it can become either testes or ovary tissue depending on which genes act next.

The X (and Y) files

Hermann Henking discovered the X chromosome while studying insects in the early 1890s. He wasn't quite sure what the lonely, unpaired structure did, but it seemed different from the rest of the chromosomes he was looking at. So rather than assign it a number (chromosomes are generally numbered according to size, largest to smallest), he called it X. In the early 1900s, Clarence McClung decided, rightly, that Henking's X was actually a chromosome, but he wasn't quite sure of its role. McClung started calling X the *accessory chromosome*. At the time, what we know as the Y chromosome carried the cumbersome moniker of *small ideochromosome*. The prefix *ideo-* means "unknown" — in other words, McClung and other geneticists of the time had no idea what the little Y guy was for.

Edmund Wilson discovered XX-XY sex determination in insects in 1905 (independent of Nettie Stevens, who accomplished the same feat that year). Wilson seems to have had the honor of naming the Y chromosome. According to three genetics historians I consulted on the topic, Wilson first used the name Y in 1909. The Y designation was in no way romantic — it was just convenient shorthand. The new name caught on rapidly, and by 1914 or so, all geneticists were calling the two sex chromosomes X and Y.

If the embryo has at least one X and lacks a Y chromosome, two genes work together to give the embryo the female phenotype. The first gene, called *DAX1,* is on the X chromosome. The second gene, *WNT4,* is on chromosome 1. Together, these genes stimulate the development of ovary tissue. The ovary tissue excretes the hormone estrogen, which turns on other genes that control the development of the remaining female reproductive structures.

The not very significant Y

In comparison to X, the Y chromosome is scrawny, antisocial, and surprisingly expendable. Y contains between 70 and 300 genes along its 50-million base pair length and is generally considered the smallest and least gene-rich human chromosome. Most of Y doesn't seem to code for any genes at all; slightly over half the Y chromosome is junk DNA. Individuals with only one X and no Y can survive the condition (known as Turner syndrome; see the section "Sex-Determination Disorders in Humans" later in this chapter), demonstrating that Y supports no genes required for survival. Almost all the genes Y has are involved in male sex determination and sexual function.

Unlike the other chromosomes, most of Y doesn't recombine during meiosis (see Chapter 2 for details) because Y is so different from X — it has only small regions near the *telomeres* (the tips of chromosomes) that allow X and Y to pair during meiosis. Pairs of human chromosomes are considered *homologous,* meaning the members of each pair are identical in structure and shape and contain similar (although not identical) genetic information.

X and Y aren't homologous — they're different in size and shape and carry completely different sets of genes. Homologous autosomes can freely swap information during meiosis (a process referred to as *crossing-over*), but X and Y don't share enough information to allow crossing-over to occur. X and Y do pair up as if they were homologous so that the right number of chromosomes gets parsed out during meiosis.

Because Y doesn't recombine with other chromosomes, it's unusually good for tracing how men have traveled and settled around the world. The Y chromosome is even helping to clarify British history. For centuries, people have believed that Anglo-Saxons conquered Britain and more or less ran everyone else out. In a 2003 survey of over 1,700 British men, however, geneticists found evidence that different parts of the British Isles have differing paternal histories reflecting a complex and rich history of invasions, immigration, and intermarriage.

The most important of Y's genes is *SRY,* the Sex-determining Region Y gene, which was discovered in 1990. The *SRY* gene is what makes men. *SRY* codes for a mere 204 amino acids (flip to Chapter 10 for how the genetic code works to make proteins from amino acids). Unlike most genes (and most of Y, for that matter), *SRY* is junk-free — it contains no *introns* (sequences that interrupt the expressed part of genes; see Chapter 9 for a full description).

SRY's most important function is starting the development of testes. Embryos that have at least one Y chromosome differentiate into males when the *SRY* gene is turned on during week seven of development. *SRY* acts with at least one other gene (on chromosome 17) to stimulate the expression of the male phenotype in the form of testes. The testes themselves secrete testosterone, the hormone responsible for the expression of most traits belonging to males. (To find out how gene expression works, turn to Chapter 11.)

Sex determination in other organisms

In mammals, sex determination is directed by the presence of sex chromosomes that turn on the appropriate genes to make male or female phenotypes. In most other organisms, however, sex determination is highly variable. This section looks at how various arrangements of chromosomes, genes, and even temperature affect the determination of sex.

Insects

When geneticists first began studying chromosomes in the early 1900s, insects were the organisms of choice. Grasshopper, beetle, and especially fruit fly chromosomes were carefully stained and studied under microscopes (check out Chapter 15 for how geneticists study chromosomes). Much of what we now know about chromosomes in general and sex determination in particular comes from the work of these early geneticists.

In 1901, Clarence McClung determined that female grasshoppers had two X chromosomes, but males had only one (take a look at the sidebar "The X (and Y) files" for more about McClung's role in discovering the sex chromosomes). This arrangement, now known as XX-XO, with the O representing a lack of a chromosome, occurs in many insects. For these organisms, the number of X chromosomes in relation to the autosomal chromosomes determines maleness or femaleness. Two doses of X produce a female. One X produces a male.

In the XX-XO system, females (XX) are *homogametic,* which means that every gamete (in this case, eggs) that the individual produces has the same set of chromosomes composed of one of each autosome and one X. Males (XO) are *heterogametic;* their sperm can come in two different types. Half of a male's gametes have one set of autosomes and an X; the other half have one set of autosomes and no sex chromosome at all. This imbalance in the number of chromosomes is what determines sex for XX-XO organisms.

A similar situation occurs in fruit flies. Male fruit flies are XY, but the Y doesn't have any sex-determining genes on it. Instead, sex is determined by the number of X chromosomes compared to the number of sets of autosomes. The number of X chromosomes an individual has is divided by the number of sets of autosomes (sometimes referred to as the haploid number, *n*; see Chapter 2). This equation is the X to autosome (A) ratio, or X:A ratio. If the X:A ratio is ½ or less, the individual is male. For example, an XX fly with two sets of autosomes would yield a ratio of 1 (2 divided by 2) and would be female. An XY fly with two sets of autosomes would yield a ratio of ½ (1 divided by 2) and would be male.

Bees and wasps have no sex chromosomes at all. Their sex is determined by whether the individual is *diploid* (with paired chromosomes) or *haploid* (with a single set of chromosomes). Females develop from fertilized eggs and are diploid. Males develop from unfertilized eggs and are therefore haploid.

Birds

Like humans, birds have two sex chromosomes: Z and W. Female birds are ZW, and males are ZZ. Sex determination in birds isn't well understood; two genes, one on the Z and the other on the W, both seem to play roles in whether an individual becomes male or female. The Z-linked gene suggests that, like the XX-XO system in insects (see the preceding section), the number of Z chromosomes may help determine sex (but with reversed results from XX-XO). On the other hand, the W-linked gene suggests the existence of a "female-determining" gene. The chicken genome sequence (see Chapter 8 for the scoop) will provide critical information for geneticists to learn how sex is determined in birds. (Sex determination for some birdlike animals is even more complex; check out the hard-to-believe story of the platypus in Chapter 24.)

Nature's gender benders

Some organisms have *location-dependent* sex determination, meaning the organism becomes male or female depending on where it ends up. Take the slipper limpet, for example. Slipper limpets (otherwise known by their highly suggestive scientific name of *Crepidula fornicata*) have concave, unpaired shells and cling to rocks in shallow seawater environments. (Basically, they look like half of an oyster.) All young slipper limpets start out as male, but a male can become female as a result of his (soon to be her) circumstances. If a young slipper limpet settles on bare rock, it becomes female. If a male settles on top of another male, the one on the bottom becomes a female to accommodate the new circumstances. If a male is removed from the top of a pile and placed on bare rock, he becomes a she and awaits the arrival of a male. After an individual becomes female, she's stuck with the change and is a female from then on.

Some fish also change sex depending on their locations or their social situations. Blue-headed wrasse, large reef fish familiar to many scuba divers, change into females if a male is present. If no male is around, or if the local male disappears, large females change sex to become males. The fish's brain and nervous system control its ability to switch from one sex to another. An organ in the brain called the *hypothalamus* (you have one, too, by the way) regulates sex hormones and controls growth of the needed reproductive tissues.

To add to the list of the truly bizarre, a parasitic critter that lives inside certain fish has an unusual way of changing gender: cannibalism. When a male *Ichthyoxenus fushanensis,* which is a sort of parasitic pill bug (you may know the isopod as a roly-poly), eats a female (or vice versa), the diner changes sex — that is, he becomes a she. In the case of the isopod, the sex change is a form of hermaphroditism where the genders are expressed sequentially and in response to some change in the environment or diet.

Reptiles

Sex chromosomes determine the sex of most reptiles (like snakes and lizards). However, the sex of most turtles and all crocodiles and alligators is determined by the temperature the eggs experience during incubation. Female turtles and crocodilians dig nests and bury their eggs in the ground. Females usually choose nest sites in open areas likely to receive a lot of sunlight. Female turtles don't bother to guard their eggs; they lay 'em and forget 'em. Alligators and crocodiles, on the other hand, guard their nests (quite aggressively, as I can personally attest) but let the warmth of the sun do the work.

In turtles, lower temperatures (78–82 degrees Fahrenheit) produce all males. At temperatures over 86 degrees, all eggs become females. Intermediate temperatures produce both sexes. Male alligators, on the other hand, are produced only at intermediate temperatures (around 91 degrees). Cooler conditions (84–88 degrees) produce only females; really warm temperatures (95 degrees) produce all females also.

An enzyme called *aromatase* seems to be the key player in organisms with temperature-dependent sex determination. Aromatase converts testosterone into estrogen. When estrogen levels are high, the embryo becomes a female. When estrogen levels are low, the embryo becomes a male. Aromatase activity varies with temperature. In some turtles, for example, aromatase is essentially inactive at 77 degrees, and all eggs in that environment hatch as males. When temperatures around the eggs get to 86 degrees, aromatase activity increases dramatically, and all the eggs become females.

Sex-Determination Disorders in Humans

Homologous chromosomes line up and part company during the first phase of meiosis, which I explain in Chapter 2. The dividing up of chromosome pairs ensures that each gamete gets only one copy of each chromosome, and thus that zygotes (created from the fusion of gametes; see Chapter 2) have one pair of each chromosome without odd copies thrown in. But sometimes, mistakes occur. Xs or Ys can get left out, or extra copies can remain. These chromosomal delivery errors are caused by *nondisjunction,* which results when chromosomes fail to segregate normally during meiosis. (Chapter 15 has more information about nondisjunction and other chromosome disorders.)

Extra chromosomes can create all sorts of developmental problems. In organisms that have chromosomal sex determination, like humans, male organisms normally have only one X, giving them one copy of each gene on the X and allowing some genes on the X chromosome to act like dominant genes when, in fact, they're recessive (take a look ahead at "X-linked disorders" for more). Female organisms have to cope with two copies, or doses, of the X chromosome and its attendant genes. If both copies of a female's X were active, she'd get twice as much X-linked gene product as a male. (*X-linked* means any and all genes on the X chromosome.) The extra protein produced by two copies of the gene acting at once derails normal development. The solution to this problem is a process called *dosage compensation,* when the amount of gene product is equalized in both sexes.

Dosage compensation is achieved in one of two ways:

- ✔ The organism increases gene expression on the X to get a double dose for males. This is what happens in fruit flies, for example.
- ✔ The female inactivates essentially all the genes on one X to get a "half" dose of gene expression.

Both methods equalize the amount of gene product produced by each sex. In humans, dosage compensation is achieved by *X inactivation;* one X chromosome is permanently and irreversibly turned off in every cell of a female's body.

X inactivation in humans is controlled by a single gene, called *XIST* (for *X Inactive-Specific Transcript*), that lies on the X chromosome. When a female zygote starts to develop, it goes through many rounds of cell division. When the zygote gets to be a little over 16 cells in size, X inactivation takes place. The *XIST* gene gets turned on and goes through the normal process of transcription (covered in Chapter 9). The RNA (a close cousin of DNA; see Chapter 9 to learn more) produced when *XIST* is transcribed isn't translated into protein (see Chapter 10 for how translation works and what it does). Instead, the *XIST* transcript binds directly to one of the X chromosomes to inactivate its genes (much like RNA interference; see Chapter 11 for the details).

X inactivation causes the entire inactivated chromosome to change form; it becomes highly condensed and genetically inert. Highly condensed chromosomes are easy for geneticists to spot because they soak up a lot of dye (see Chapter 15 for how geneticists study chromosomes using dyes). Murray Barr was the first person to observe the highly condensed, inactivated X chromosomes in mammals. Therefore, these inactivated chromosomes are called *Barr bodies*.

You should remember two very important things about X inactivation:

- In humans, X inactivation is random. Only one X remains turned on, but which X remains on is completely up to chance.
- If more than two Xs are present, only one remains completely active.

The ultimate result of X inactivation is that the tissues that arise from each embryonic cell have a "different" X. Because females get one X from their father and the other from their mother, their Xs are likely to carry different alleles of the same genes. Therefore, their tissues may express different phenotypes depending on which X (Mom's or Dad's) remains active. This random expression of X chromosomes is best illustrated in cats.

Calico and tortoiseshell cats both have patchy-colored fur (often orange and black, but other combinations are possible). The genes that control these fur colors are on the X chromosomes. Male cats are usually all one color because they always have only one active X chromosome (and are XY). Females (XX), on the other hand, also have one active X chromosome, but the identity of the active X (maternal or paternal) varies over the cat's body. Therefore, calico females get a patchy distribution of color depending on which X is active (that is, as long as her parents had different alleles on their Xs). If you have a calico male cat, he possesses an extra X and has the genotype XXY. XXY cats have normal phenotypes. Unlike cats, humans with extra sex chromosomes have a variety of health problems, which I summarize later in this chapter.

Extra X's

Both males and females can have multiple X chromosomes, each with different genetic and phenotypic consequences. When females have extra X chromosomes, the condition is referred to as *Poly-X* (*poly* meaning "many"). Poly-X females tend to be taller than average and often have a thin build. Most Poly-X women develop normally and experience normal puberty, menstruation, and fertility. Rarely, XXX (referred to as *Triplo-X*) females have mental retardation; the severity of mental retardation and other health problems experienced by Poly-X females increases with the number of extra Xs. About one in every 1,000 girls is XXX.

Males with multiple X chromosomes are affected with *Klinefelter syndrome*. Roughly one in every 500 boys is XXY. Most often, males with Klinefelter are XXY, but some males have as many as four extra X chromosomes. Like females, males affected by Klinefelter undergo X inactivation so that only one X chromosome is active. However, the extra X genes act in the embryo before X inactivation takes place. These extra doses of X genes are responsible for the phenotype of Klinefelter. Generally, males with Klinefelter are taller than average and have impaired fertility (usually they're sterile). Men with Klinefelter often have reduced secondary sexual characteristics (such as less facial hair) and sometimes have some breast enlargement due to impaired production of testosterone.

For additional information and to find contacts in your area, contact Klinefelter Syndrome and Associates at 1-888-999-9428 (www.genetic.org) or the American Association for Klinefelter Syndrome Information and Support at 1-888-466-5747 (www.aaksis.org).

Extra Y's

Occasionally, human males have two or more Y chromosomes and one X chromosome. Most XYY men have a normal male phenotype, but they're often taller and, as children, grow a bit faster than their XY peers. Studies conducted during the 1960s and 1970s indicated that XYY men were more prone to criminal activity than XY men. Since then, findings have documented learning disabilities (XYY boys may start talking later than XY boys), but it seems that XYY males are no more likely to commit crimes than XY males.

One X and no Y

In some cases, individuals end up with one X chromosome. Such individuals have *Turner syndrome* and are female. Often, affected persons never undergo puberty and don't acquire secondary sex characteristics of adult women (namely, breast development and menstruation), and they tend to have short stature. In most other ways, girls and women with Turner syndrome are completely normal. Occasionally, however, they have kidney or heart defects. Turner syndrome (also referred to as *monosomy X,* meaning only one X is present) affects about one in 2,500 girls.

For additional information and to find contacts in your area, contact the Turner Syndrome Society of the United States at 1-800-365-9944 (or online at `www.turner-syndrome-us.org`) or the Turner Syndrome Society of Canada at 1-800-465-6744 (`www.turnersyndrome.ca`).

Found on Sex Chromosomes: Sex-linked Inheritance

Sex not only controls an organism's reproductive options; it also has a lot to do with which genes are expressed and how. *Sex-linked genes* are ones that are actually located on the sex chromosomes themselves. Some traits are truly X-linked (such as hemophilia) or Y-linked (such as hairy ears). Other traits are expressed differently in males and females even though the genes that control the traits are located on nonsex chromosomes. This section explains how sex influences (and sometimes controls) the phenotypes of various genetic conditions.

X-linked disorders

Genes on the X chromosome control X-linked traits. In 1910, Thomas H. Morgan discovered X-linked inheritance while studying fruit flies. Morgan's observations made him doubt the validity of Mendelian inheritance (see Chapter 3). His skepticism about Mendelian inheritance stemmed from the fact that he kept getting unexpected phenotypic ratios when he crossed red- and white-eyed flies. He thought the trait of white eyes was simply recessive, but when he crossed red-eyed females with white-eyed males, he got all red-eyed flies — the exact result you'd expect from a monohybrid cross. The F2 generation showed the expected 3:1 ratio, too.

But when Morgan crossed white-eyed females with red-eyed males, all the expected relationships fell apart. The F1 generation had a 1:1 ratio of white- to red-eyed flies. In the F2, the phenotypic ratio of white-eyed to red-eyed flies was also 1:1 — not at all what Mendel would have predicted. Morgan was flustered until he looked at which sex showed which phenotype.

In Morgan's F1 offspring from his white-eyed mothers and red-eyed fathers, all the sons were white-eyed (see Figure 5-2). Daughters of white-eyed females were red-eyed. In the F2, Morgan got equal numbers of white- and red-eyed males and females.

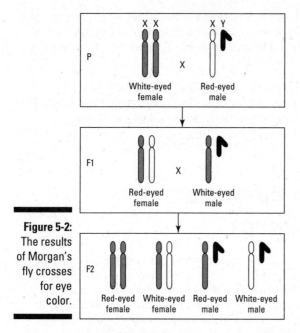

Figure 5-2:
The results
of Morgan's
fly crosses
for eye
color.

Morgan was well aware of the work on sex chromosomes conducted by
Nettie Stevens and Edmund Wilson in 1905, and he knew that fruit flies have
XX-XY sex chromosomes. Morgan and his students examined the phenotypes
of 13 million fruit flies to confirm that the gene for eye color was located on
the X chromosome. (The next time you see a fruit fly in your kitchen, imagine
looking through a microscope long enough to examine 13 million flies!)

As it turns out, the gene for white eye color in fruit flies is recessive. The only
time it's expressed in females is when it's homozygous. Males, on the other
hand, show the trait when they have only one copy of the X-linked gene. For
all X-linked recessive traits, the gene acts like a dominant gene when it's in
the *hemizygous* (one copy) state. Any male inheriting the affected X chromosome shows the trait as if it were present in two copies (X-linked dominant
disorders also occur; see Chapter 12 for the details).

In humans, X-linked recessive disorders rarely show up in females. Instead,
X-linked recessive traits affect sons of women who are carriers. To see the
distribution of X-linked recessive disorders, check out the family tree for the
royal families of Europe in Chapter 12. Queen Victoria was apparently a carrier for the X-linked gene that causes hemophilia. None of Queen Victoria's
ancestors appears to have had hemophilia; geneticists think that the mutation originated with Queen Victoria herself (see Chapter 13 for more about
spontaneous mutations like these). Queen Victoria had one son with hemophilia, and two of her daughters were carriers.

Sex-limited traits

Sex-limited traits are inherited in the normal autosomal fashion but are never expressed in one sex, regardless of whether the gene is heterozygous or homozygous. Such traits are said to have 100 percent penetrance in one sex and zero penetrance in the other. (*Penetrance* is the probability that an individual having a dominant allele will show its effects; see Chapter 4 for more.) Traits such as color differences between male and female birds are sex-limited; both males and females inherit the genes for color, but the genes are expressed only in one sex (usually the male). In mammals, both males and females possess the genes necessary for milk production, but only females express these genes, which are controlled by hormone levels in the female's body (see Chapter 11 for more about how gene expression is controlled).

One trait in humans that's male-limited is precocious puberty. The corresponding gene, located on chromosome 2, causes boys to undergo the changes associated with teenage years, such as a deeper voice and beard and body hair growth, at very early ages (sometimes as young as 3 years of age). The allele responsible for precocious puberty acts as an autosomal dominant, expressed only in males. Females, regardless of genotype, never exhibit this kind of precocious puberty.

Sex-influenced traits

Sex-influenced traits are coded by genes on autosomes, but the phenotype depends on the sex of the individual carrying the affected gene. Sex-influenced traits come down to the issue of penetrance: The traits are more penetrant in males than females. Horns, hair, and other traits that make male organisms look different from females are usually sex-influenced traits.

In humans, male-pattern baldness is a sex-influenced trait. The gene credited with male hair loss is on chromosome 15. Baldness is autosomal dominant in men, and women only show the phenotype of hair loss when they're homozygous for the gene. The gene for male-pattern baldness has also been implicated in polycystic ovary disease in women. Women with polycystic ovary disease experience reduced fertility and other disorders of the reproductive system. The gene seems to act as an autosomal dominant for ovarian disease in women, much as it does for male-pattern baldness in men, so women with ovarian disease are usually heterozygous for the condition (and thus, not bald).

Y-linked traits

The Y chromosome carries few genes, and the genes it does carry are all related to male sex determination. Therefore, most of the Y-linked traits discovered so far have something to do with male sexual function and fertility. As you may expect, Y-linked traits are passed strictly from father to son. All Y-linked traits are expressed, because the Y is hemizygous (having one copy) and therefore has no other chromosome to offset gene expression. The amount of penetrance and expressivity that Y-linked traits show varies (see Chapter 4 for more details about penetrance and expressivity of autosomal dominant traits).

One trait that seems to be Y-linked but isn't related to sexual function is hairy ears. Men with hairy ears grow varying amounts of hair on their outer ears or from the ear canals. The trait appears to be incompletely penetrant, meaning not all sons of hairy-eared fathers show the trait. Hairy ears also show variable expressivity from very hairy to only a few stray hairs. Aren't you glad that geneticists have focused the powers of science at their disposal on making such important discoveries? Check out the section "The not very significant Y" earlier in this chapter for a rundown of other Y-linked genes in humans and other mammals.

Part II

DNA: The Genetic Material

The 5th Wave   By Rich Tennant

In this part . . .

All life on earth depends on the essentially iconic double helix that holds all the genetic information of each and every individual. The physical and chemical makeup of DNA is responsible for DNA's massive storage capacity and controls how it's copied and how its message is passed on.

In this part, I explain how DNA is put together and how the messages are read and ultimately expressed as the traits of the organisms you see every day. The genetic code relies on DNA's close cousin, RNA, to carry the important messages of genes. The ultimate fate of DNA's messages is to create proteins, the building blocks of life. The following chapters tell you all about how DNA's blueprint is assembled from start to finish.

Chapter 6

DNA: The Basis of Life

● ●

In This Chapter

▶ Identifying the chemical components of DNA

▶ Understanding the structure of the double helix

▶ Looking at different DNA varieties

▶ Chronicling DNA's scientific history

● ●

*1*t's time to meet the star of the genetics show: *deoxyribonucleic acid,* otherwise known as DNA. If the title of this chapter hasn't impressed upon you the importance and magnitude of those three little letters, consider that DNA is also referred to as "*the* genetic material" or "*the* molecule of heredity." And you thought your title was impressive!

Every living thing on earth, from the smallest bacteria to the largest whale, uses DNA to store genetic information and transmits that info from one generation to the next; a copy of some (or all) of every creature's DNA is passed on to its offspring. The developing organism then uses DNA as a blueprint to make all its body parts. (Some non-living things use DNA to transmit information, too; see the nearby sidebar "DNA and the undead: The world of viruses" for details.)

To get an idea of how much information DNA stores, think about how complex your body is. You have hundreds of kinds of tissues that all perform different functions. It takes a lot of DNA to catalog all that. (See the section "Discovering DNA" later in this chapter to find out how scientists learned that DNA is the genetic material of all known life forms.)

The structure of DNA provides a simple way for the molecule to copy itself (see Chapter 7) and protects genetic messages from getting garbled (see Chapter 13). That structure is at the heart of forensic methods used to solve crimes, too (see Chapter 18). But before you can start exploring genetic information and applications of DNA, you need to have a handle on its chemical makeup and physical structure. In this chapter, I explore the essential makeup of DNA and the various sorts of DNA present in living things.

DNA and the undead: The world of viruses

Viruses contain DNA, but they aren't considered living things. To reproduce, a virus must attach itself to a living cell. As soon as the virus finds a host cell, the virus injects its DNA into the cell and forces that cell to reproduce the virus. A virus can't grow without stealing energy from a living cell, and it can't move from one organism to another on its own. Although viruses come in all sorts of fabulous shapes, they don't have all the components that cells do; in general, a virus is just DNA surrounded by a protein shell. So a virus isn't alive, but it's not quite dead either. Creepy, huh?

Deconstructing the Double Helix

If you're like most folks, when you think of DNA, you think of a double helix. But DNA isn't just a double helix; it's a *huge* molecule — so huge that it's called a *macromolecule*. It can even be seen with the naked eye! (Check out the nearby sidebar "Molecular madness: Extracting DNA at home" for an experiment you can do to see actual DNA.) If you were to lay out, end to end, all the DNA from just one of your cells, the line would be a little over 6 feet long! You have roughly 100,000,000,000,000 cells in your body (that's 100 trillion, if you don't feel like counting zeros). Put another way, laid out altogether, the DNA in your body would easily stretch to the sun and back — nearly 100 times!

You're probably wondering how a huge DNA molecule can fit into a teeny tiny cell so small that you can't see it with the naked eye. Here's how: DNA is tightly packed in a process called *supercoiling*. Much like a phone cord that's been twisted around and around on itself, supercoiling takes DNA and wraps it around proteins to form *nucleosomes*. Other proteins, called *histones,* hold the coils together. The nuclesomes and histones together form a structure similar to beads on a string. The whole "necklace" twists around itself so tightly that over 6 feet of DNA is compressed into only a few thousandths of an inch.

Although the idea of a DNA path to the sun works great for visualizing the size of the DNA molecule, an organism's DNA usually doesn't exist as one long piece. Rather, strands of DNA are divided into *chromosomes,* which are relatively short pieces. (I introduce chromosomes in Chapter 2 and discuss related disorders in Chapter 15.) In humans and all other *eukaryotes* (organisms whose cells have nuclei; see Chapter 2 for more), a full set of chromosomes is stored in the nucleus of each cell. That means that practically every cell contains a complete set of instructions to build the entire organism! The instructions are packaged as *genes.* A gene determines exactly how a specific trait will be expressed. Genes and how they work are topics I discuss in detail in Chapter 11.

Molecular madness: Extracting DNA at home

Using this simple recipe, you can see DNA right in the comfort of your own home! You need a strawberry, salt, water, two clear jars or juice glasses, a sandwich bag, a measuring cup, a white coffee filter, clear liquid soap, and rubbing alcohol. (Other foods such as onions, bananas, kiwis, and tomatoes also work well if strawberries are unavailable.) After you've gathered these ingredients, follow these steps:

1. **Put slightly less than ⅜ cup of water into the measuring cup. Add ¼ teaspoon of salt and enough clear liquid soap to make ⅜ cup of liquid altogether. Stir gently until the salt dissolves into the solution.**

   The salt provides sodium ions needed for the chemical reaction that allows you to see the DNA in Step 6. The soap causes the cell walls to burst, freeing the DNA inside.

2. **Remove the stem from the strawberry, place the strawberry into the sandwich bag, and seal the bag. Mash the strawberry thoroughly until completely pulverized (I roll a juice glass repeatedly over my strawberry to pulverize it). Make sure you don't puncture the bag.**

3. **Add 2 teaspoons of the liquid soap-salt solution to the bag with the strawberry, and then reseal the bag. Mix gently by** compressing the bag or rocking the bag back and forth for at least 45 seconds to one minute.

4. **Pour the strawberry mixture through the coffee filter into a clean jar. Let the mixture drain into the jar for 10 minutes.**

   Straining gets rid of most of the cellular debris (a fancy word for gunk) and leaves behind the DNA in the clean solution.

5. **While the strawberry mixture is draining, pour ¼ cup of rubbing alcohol into a clean jar and put the jar in the freezer. After 10 minutes have elapsed, discard the coffee filter and pulverized strawberry remnants. Put the jar with the cold alcohol on a flat surface where it will be undisturbed, and pour the strained strawberry liquid into the alcohol.**

6. **Let the jar sit for at least 5 minutes, and then check out the result of your DNA experiment. The cloudy substance that forms in the alcohol layer is the DNA from the strawberry. The cold alcohol helps strip the water molecules from the outside of the DNA molecule, causing the molecule to collapse on itself and "fall out" of the solution.**

Cells with nuclei are found only in eukaryotes; however, not every eukaryotic cell has a nucleus. For example, humans are eukaryotes, but human red blood cells don't have nuclei. For more on cells, flip to Chapter 2.

The tutorial offered at www.umass.edu/molvis/tutorials/dna/ provides an excellent complement to the information on the structure of DNA I cover in this section. You can access incredible, interactive views of precisely how DNA is put together to form the double helix. A click-and-drag feature allows you to turn the molecule in any direction to better understand the structure of the genetic material, to highlight different parts of the molecule, and to see exactly how all the parts fit together.

Chemical ingredients of DNA

DNA is a remarkably durable molecule; it can be stored in ice or in a fossilized bone for thousands of years. DNA can even stay in one piece for as long as 100,000 years under the right conditions. This durability is why scientists can recover DNA from 14,000-year-old mammoths and learn that the mammoth is most closely related to today's Asian elephants. (Scientists have recovered ancient DNA from an amazing variety of organisms — check out the sidebar "Still around after all these years: Durable DNA" for more.) The root of DNA's extreme durability lies in its chemical and structural makeup.

CASE STUDY

Still around after all these years: Durable DNA

When an organism dies, it starts to decay and its DNA starts to break down (for DNA, this means breaking into smaller and smaller pieces). But if a dead organism dries out or freezes shortly after death, decay slows down or even stops. Because of this kind of interference with decay, scientists have been able to recover DNA from animals and humans that roamed the earth as many as 100,000 years ago. This recovered DNA tells scientists a lot about life and the conditions of the world long ago. But even this very durable molecule has its limits — about a million years or so.

In 1991, hikers in the Italian Alps discovered a human body frozen in a glacier. As the glacier melted, the retreating ice left behind a secret concealed for over 5,000 years: an ancient human. The Ice Man, renamed Otzi, has yielded amazing insight into what life was like in northern Italy thousands of years ago. Scientists have recovered DNA from this lonely shepherd, his clothing, and even the food in his stomach. Apparently, red deer and ibex meat were part of his last meal. His food was dusted with pollen from nearby trees, so even the forest he walked through can be identified!

By analyzing Otzi's mitochondrial (mt) DNA, which he inherited from his mother (see the "Mitochondrial DNA" section later in this chapter), scientists discovered that he wasn't related to any modern European population studied so far. A team of investigators from Australia, led by the late Thomas Loy, examined blood found on Otzi's clothing and possessions. Like modern forensic scientists, Loy's team determined that four different people's DNA fingerprints were present, in addition to Otzi's own (to find out how DNA fingerprints are used to solve modern crimes, check out Chapter 18). The team found blood from two different people on Otzi's arrow, a third person's blood on his knife, and a fourth person's blood on his clothing. These findings led people to speculate that he was involved in a fight shortly before he died.

Otzi isn't the only ancient human whose DNA scientists are analyzing. Neandertals were humans that roamed the earth up to about 30,000 years ago (give or take several centuries). Using 38,000-year-old mtDNA, researchers have discovered that Neandertals had a substantially different mtDNA profile than modern humans, suggesting that while modern humans and Neandertals lived at the same time, they probably didn't interbreed (or if they did, none of the descendants survived to be represented in human populations now). In addition, Neandertals were lactose-intolerant; they lacked the gene that codes for the enzyme that breaks down lactose (a sugar present in milk). Neandertals probably were able to speak much as we do — they carried a version of the gene associated with human speech.

REMEMBER

Chemically, DNA is really simple. It's made of three components: nitrogen-rich bases, deoxyribose sugars, and phosphates. The three components, which I explain in the following sections, combine to form a *nucleotide* (see the section "Assembling the double helix: The structure of DNA" later in this chapter). Thousands of nucleotides come together in pairs to form a single molecule of DNA.

Covering the bases

Each DNA molecule contains thousands of copies of four specific nitrogen-rich bases:

- ✔ Adenine (A)

- ✔ Guanine (G)

- ✔ Cytosine (C)

- ✔ Thymine (T)

As you can see in Figure 6-1, the bases are comprised of carbon (C), hydrogen (H), nitrogen (N), and oxygen (O) atoms.

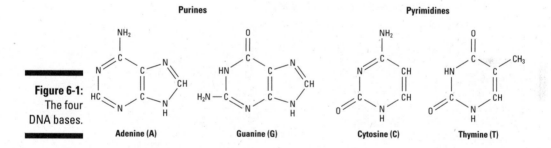

Figure 6-1: The four DNA bases.

The four bases come in two flavors:

- ✔ **Purines:** The two purine bases in DNA are adenine and guanine. If you were a chemist, you'd know that the word *purine* means a compound composed of two rings (check out adenine's and guanine's structures in Figure 6-1). If you're like me (not a chemist), you're likely still familiar with one common purine: caffeine.

- ✔ **Pyrimidines:** The two pyrimidine bases in DNA are cytosine and thymine. The term *pyrimidine* refers to chemicals that have a single six-sided ring structure (see cytosine's and thymine's structures in Figure 6-1).

REMEMBER

Because they're rings, all four bases are flat molecules. And as flat molecules, they're able to stack up in DNA much like a stack of coins. The stacking arrangement accomplishes two things: It makes the molecule both compact and very strong.

It's been my experience that students and other folks get confused by spatial concepts where DNA is concerned. To see the chemical structures more easily, DNA is often drawn as if it were a flattened ladder. But in its true state, DNA isn't flat — it's three-dimensional. Because DNA is arranged in strands, it's also linear. One way to think about this structure is to look at a phone cord (that is, if you can find a phone that isn't cordless). A phone cord spirals in three dimensions, yet it's linear (rope-like) in form. That's sort of the shape DNA has, too.

The bases carry the information of DNA, but they can't bond together by themselves. Two more ingredients are needed: a special kind of sugar and a phosphate.

Adding a spoonful of sugar and a little phosphate

To make a complete nucleotide (thousands of which combine to make one DNA molecule), the bases must attach to deoxyribose and a phosphate molecule. *Deoxyribose* is ribose sugar that has lost one of its oxygen atoms. When your body breaks down *adenosine triphosphate* (ATP), the molecule your body uses to power your cells, ribose is released with a phosphate molecule still attached to it. Ribose loses an oxygen atom to become deoxyribose (see Figure 6-2) and holds onto its phosphate molecule, which is needed to transform a lone base into a nucleotide.

Figure 6-2:
The chemical structure of ribose and deoxyribose.

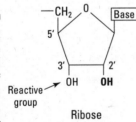

Ribose is the precursor for deoxyribose and is the chemical basis for RNA (see Chapter 9). The only difference between ribose and deoxyribose sugars is the presence or absence of an oxygen atom at the 2' site.

Chemical structures are numbered so you can keep track of where atoms, branches, chains, and rings appear. On ribose sugars, numbers are followed by an apostrophe (') to indicate the designation "prime." The addition of "prime" prevents confusion with numbered sites on other molecules that bond with ribose.

Deoxy- means that an oxygen atom is missing from the sugar molecule and defines the *D* in DNA. As an added touch, some authors write "2-" before the "deoxy-" to indicate which site lacks the oxygen — the number 2 site, in this case. The OH group at the 3' site of both ribose and deoxyribose is a *reactive group.* That means the oxygen atom at that site is free to interact chemically with other molecules.

Assembling the double helix: The structure of DNA

Nucleotides are the true building blocks of DNA. In Figure 6-3, you see the three components of a single nucleotide: one deoxyribose sugar, one phosphate, and one of the four bases. (Flip back to "Chemical ingredients of DNA" for the details of these components.) To make a complete DNA molecule, single nucleotides join to make chains that come together as matched pairs and form long double strands. This section walks you through the assembly process. To make the structure of DNA easier to understand, I start with how a single strand is put together.

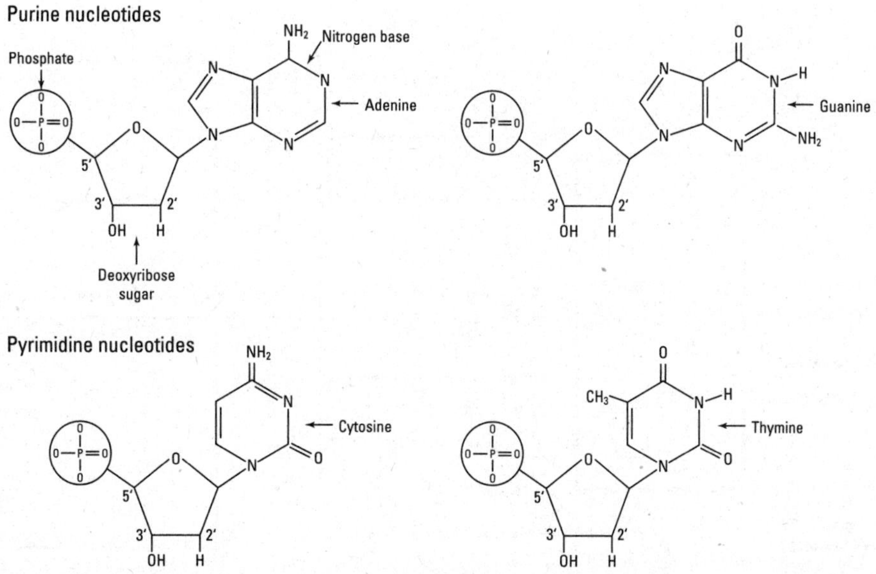

Purine nucleotides

Pyrimidine nucleotides

Figure 6-3: Chemical structures of the four nucleotides present in DNA.

DNA normally exists as a double-stranded molecule. In living things, new DNA strands are *always* put together using a preexisting strand as a pattern (see Chapter 7).

Starting with one: Weaving a single strand

Hundreds of thousands of nucleotides link together to form a strand of DNA, but they don't hook up haphazardly. Nucleotides are a bit like coins in that they have two "sides" — a phosphate side and a sugar side. Nucleotides can only make a connection by joining phosphates to sugars. The bases wind up parallel to each other (stacked like coins), and the sugars and phosphates run perpendicular to the stack of bases. A long strand of nucleotides put together in this way is called a *polynucleotide* strand (*poly* meaning "many"). In Figure 6-4, you can see how the nucleotides join together; a single strand would comprise one-half of the two-sided molecule (the chain of sugars, phosphates, and one of the pair of bases).

Because of the way the chemical structures are numbered, DNA has numbered "ends." The phosphate end is referred to as the 5' (5-prime) end, and the sugar end is referred to as the 3' (3-prime) end. (If you missed the discussion of how the chemical structure of deoxyribose is numbered, check out the earlier section "Adding a spoonful of sugar and a little phosphate.") The bonds between a phosphate and two sugar molecules in a nucleotide strand are collectively called a *phosphodiester bond.* This is a fancy way of saying that two sugars are linked together by a phosphate in between.

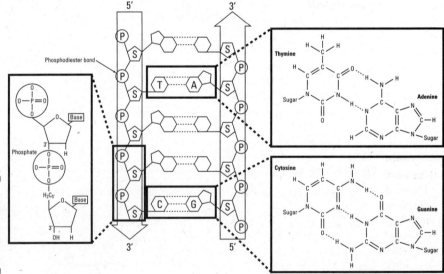

Figure 6-4:
The chemical structures of DNA.

After they're formed, strands of DNA don't enjoy being single; they're always looking for a match. The arrangement in which strands of DNA match up is very, very important. A number of rules dictate how two lonely strands of DNA find their perfect matches and eventually form the star of the show, the molecule you've been waiting for — the double helix.

Doubling up: Adding the second strand

A complete DNA molecule has

✔ Two side-by-side polynucleotide strands twisted together

✔ Bases attached in pairs in the center of the molecule

✔ Sugars and phosphates on the outside, forming a backbone

If you were to untwist a DNA double helix and lay it flat, it would look a lot like a ladder (refer to Figure 6-4). The bases are attached to each other in the center to make the rungs, and the sugars are joined together by phosphates to form the sides of the ladder. It sounds pretty straightforward, but this ladder arrangement has some special characteristics.

If you were to separate the ladder into two polynucleotide strands, you'd see that the strands are oriented in opposite directions (shown with arrows in Figure 6-4). The locations of the sugar and the phosphate give nucleotides heads and tails, two distinct ends. (If you skipped that part, it's in the earlier section "Starting with one: Weaving a single strand.") The heads-tails (or in this case, 5'-3') orientation applies here. This head-to-tail arrangement is called *antiparallel,* which is a fancy way of saying that something is parallel and running in opposite directions. Part of the reason the strands must be oriented this way is to guarantee that the dimensions of the DNA molecule are even along its entire length. If the strands were put together in a parallel arrangement, the angles between the atoms would be all wrong, and the strands wouldn't fit together.

The molecule is guaranteed to be the same size all over because the matching bases *complement* each other, making whole pieces that are all the same size. Adenine complements thymine, and guanine complements cytosine. The bases *always* match up in this complementary fashion. Therefore, in every DNA molecule, the amount of one base is equal to the amount of its complementary base. This condition is known as *Chargaff's rules* (see the "Obeying Chargaff's rules" section later in the chapter for more on the discovery of these rules).

Why can't the bases match up in other ways? First, purines are larger than pyrimidines (see "Covering the bases" earlier in the chapter). So matching like with like would introduce irregularities in the molecule's shape. Irregularities are bad because they can cause mistakes when the molecule is copied (see Chapter 13).

An important result of the bases' complementary pairing is the way in which the strands bond to each other. Hydrogen bonds form between the base pairs. The number of bonds between the base pairs differs; G-C (guanine-cytosine) pairs have three bonds, and A-T (adenine-thymine) pairs have only two. Figure 6-4 illustrates the structure of the untwisted double helix — specifically, the bonds between base pairs. Every DNA molecule has hundreds of thousands of base pairs, and each base pair has multiple bonds, so the rungs of the ladder are very strongly bonded together.

When inside a cell, the two strands of DNA gently twist around each other like a spiral staircase (or a strand of licorice, or the stripes on a candy cane . . . anybody else have a sweet tooth?). The antiparallel arrangement of the two strands is what causes the twist. Because the strands run in opposite directions, they pull the sides of the molecule in opposite directions, causing the whole thing to twist around itself.

Most naturally occurring DNA spirals clockwise, as you can see in Figure 6-5. A full twist (or complete turn) occurs every ten base pairs or so, with the bases safely protected on the inside of the helix. The helical form is one way that the information that DNA carries is protected from damage that can result in mutation.

The helical form creates two grooves on the outside of the molecule (see Figure 6-5). The major groove actually lets the bases peep out a little, which is important when it's time to read the information DNA contains (see Chapter 10).

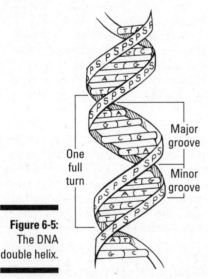

One full turn

Major groove

Minor groove

Figure 6-5:
The DNA double helix.

Because base pairs in DNA are stacked on top of each other, chemical interactions make the center of the molecule repel water. Molecules that repel water are called *hydrophobic* (Greek for "afraid of water"). The outside of the DNA molecule is just the opposite; it attracts water. The result is that the inside of the helix remains safe and dry while the outside is encased in a "shell" of water.

Here are a few additional details about DNA that you need to know:

- ✔ **A DNA strand is measured by the number of base pairs it has.**

- ✔ **The sequence of bases in DNA isn't random.** The genetic information in DNA is carried in the order of the base pairs. In fact, the genes are encoded in the base sequences. Chapter 10 explains how the sequences are read and decoded.

- ✔ **DNA uses a preexisting DNA strand as a pattern or template in the assembly process.** DNA doesn't just form on its own. The process of making a new strand of DNA using a preexisting strand is called *replication*. I cover replication in detail in Chapter 7.

Examining Different Varieties of DNA

All DNA has the same four bases, obeys the same base pairing rules, and has the same double helix structure. No matter where it's found or what function it's carrying out, DNA is DNA. That said, different sets of DNA exist within a single organism. These sets carry out different genetic functions. In this section, I explain where the various DNAs are found and describe what they do.

Nuclear DNA

Nuclear DNA is DNA in cell nuclei, and it's responsible for the majority of functions that cells carry out. Nuclear DNA carries codes for *phenotype*, the physical traits of an organism (for a review of genetics terms, see Chapter 3). Nuclear DNA is packaged into chromosomes and passed from parent to offspring (see Chapter 2). When scientists talk about sequencing the human genome, they mean human nuclear DNA. (A *genome* is a full set of genetic instructions; see Chapter 11 for more about the human genome.) The nuclear genome of humans is comprised of the DNA from all 24 chromosomes (22 autosomes plus one X and one Y; see Chapter 2 for chromosome lingo).

Mitochondrial DNA

Animals, plants, and fungi all have *mitochondria* (for a review of cell parts, turn to Chapter 2). These powerhouses of the cell come with their own DNA, which is quite different in form (and inheritance) from nuclear DNA (see the preceding section). Each *mitochondrion* (the singular word for *mitochondria*) has many molecules of mitochondrial DNA — *mtDNA,* for short.

Mighty mitochondria

Mitochondrial DNA (mtDNA) bears a strong resemblance to a bacterial DNA. The striking similarities between mitochrondria and a certain bacteria called *Rickettsia* have led scientists to believe that mitochrondria originated from *Rickettsia*. *Rickettsia* causes typhus, a flu-like disease transmitted by flea bites (the flea first bites an infected rat or mouse and then bites a person). As for the similarities, neither *Rickettsia* nor mitochondria can live outside a cellular home, both have circular DNA, and both share similar DNA sequences (see Chapter 8 for how DNA sequences are compared between organisms). Instead of being parasitic like *Rickettsia,* however, mitochondria are considered *endosymbiotic,* meaning they must be inside a cell to work *(endo-)* and they provide something good to the cell *(-symbiotic).* In this case, the something good is energy.

Because mtDNA is passed only from mother to child (see the earlier section "Mitochondrial DNA" for an explanation), scientists have compared mtDNA from people all over the world to investigate the origins of modern humans. These comparisons have led some scientists to believe that all modern humans have one particular female ancestor in common, a woman who lived on the African continent about 200,000 years ago. This hypothetical woman has been called "Mitochondrial Eve," but she wasn't the only woman of her time. There were many women, but apparently, none of their descendants survived, making Eve what scientists refer to as our "most recent common ancestor," or MRCA. Some evidence suggests that all humans are descended from a rather small population of about 100,000 individuals, meaning that all people on earth have common ancestry.

Whereas human nuclear DNA is linear, mtDNA is circular (hoop-shaped). Human mtDNA is very short (slightly less than 17,000 base pairs) and has 37 genes, which account for almost the entire mtDNA molecule. These genes control cellular metabolism — the processing of energy inside the cell.

Half of your nuclear DNA came from your mom, and the other half came from your dad (see Chapter 2 for the scoop on how meiosis divides up chromosomes). But *all* your mtDNA came from your mom. All your mom's mtDNA came from her mom, and so on. All mtDNA is passed from mother to child in the cytoplasm of the egg cell (go to Chapter 2 for cell review).

Sperm cells have essentially no cytoplasm and thus, virtually no mitochondria. Special chemicals in the egg destroy the few mitochondria that sperm do possess.

Chloroplast DNA

Plants have three sets of DNA: nuclear in the form of chromosomes, mitochondrial, and *chloroplast DNA* (cpDNA). Chloroplasts are organelles found only in plants, and they're where *photosynthesis* (the conversion of light to chemical energy) occurs. To complicate matters, plants have mitochondria (and thus mtDNA) in their chloroplasts. Like mitochondria, chloroplasts probably originated from bacteria (see the sidebar "Mighty mitochondria").

Chloroplast DNA molecules are circular and fairly large (120,000–160,000 base pairs) but only have about 120 genes. Most of those genes supply information used to carry out photosynthesis. Inheritance of cpDNA can be either maternal or paternal, and cpDNA, along with mtDNA, is transmitted to offspring in the cytoplasm of the seed.

Digging into the History of DNA

Back when Mendel was poking around his pea pods in the early 1860s (see Chapter 3), neither he nor anybody else knew about DNA. DNA was discovered in 1868, but its importance as *the* genetic material wasn't appreciated until nearly a century later. This section gives you a rundown on how DNA and its role in inheritance was revealed.

Discovering DNA

In 1868, a Swiss medical student named Johann Friedrich Miescher isolated DNA for the first time. Miescher was working with white blood cells that he obtained from the pus drained out of surgical wounds (yes, this man was dedicated to his work). Eventually, Miescher established that the substance he called *nuclein* was rich in phosphorus and was acidic. Thus, one of his students renamed the substance *nucleic acid,* a name DNA still carries today. Like Mendel's findings on the inheritance of various plant traits, Miescher's work wasn't recognized for its importance until long after his death, and it took 84 years for DNA to be recognized as *the* genetic material. Until the early 1950s, everyone was sure that protein had to be the genetic material because, with only four bases, DNA seemed too simple.

In 1928, Frederick Griffith recognized that bacteria could acquire something — he wasn't quite sure what — from each other to transform harmless bacteria into deadly bacteria (see Chapter 22 for the whole story). A team of scientists led by Oswald Avery followed up on Griffith's experiments and determined that the "transforming principle" was DNA. Even though Avery's results were solid, scientists of the time were skeptical about the significance of DNA's role in inheritance. It took another elegant set of experiments using a virus that infected bacteria to convince the scientific community that DNA was the real deal.

Alfred Chase and Martha Hershey worked with a virus called a *bacteriophage* (which means "eats bacteria," even though the virus actually ruptures the bacteria rather than eats it). Bacteriophages grab onto the bacteria's cell wall and inject something into the bacteria. At the time of Hershey and Chase's experiment, the injected substance was unidentified. The bacteriophage produces its offspring inside the cell and then bursts the cell wall open to free the viral "offspring." Offspring carry the same traits as the original attacking bacteriophage, so it was certain that whatever got injected must be the genetic material, given that most of the bacteriophage stays stuck on the outside of the cell. Hershey and Chase attached radioactive chemicals to track different parts of the bacteriophage; for example, they used sulfur to track protein, because proteins contain sulfur, and DNA was marked with phosphorus (because of the sugar-phosphate backbone). Hershey and Chase reasoned that offspring bacteriophages would get marked with one or the other, depending on which — DNA or protein — turned out to be the genetic material. The results showed that the viruses injected only DNA into the bacterial cell to infect it. All the protein stayed stuck on the outside of the bacterial cell. They published their findings in 1952, when Hershey was merely 24 years old!

Obeying Chargaff's rules

Long before Hershey and Chase published their pivotal findings, Erwin Chargaff read Oswald Avery's paper on DNA as the transforming principle (see Chapter 22) and immediately changed the focus of his entire research program. Unlike many scientists of his day, Chargaff recognized that DNA was the genetic material.

Chargaff focused his research on learning as much as he could about the chemical components of DNA. Using DNA from a wide variety of organisms, he discovered that all DNA had something in common: When DNA was broken into its component bases, the amount of guanine fluctuated wildly from one organism to another, but the amount of guanine always equaled the amount of cytosine. Likewise, in every organism he studied, the amount of adenine equaled the amount of thymine. Published in 1949, these findings are so consistent that they're called *Chargaff's rules*. Unfortunately, Chargaff was unable to realize the meaning of his own work. He knew that the ratios said something important about the structure of DNA, but he couldn't figure out what that something was. It took a pair of young scientists named Watson and Crick — Chargaff called them "two pitchmen in search of a helix" — to make the breakthrough.

Hard feelings and the helix: Franklin, Wilkins, Watson, and Crick

If you don't know the name Rosalind Franklin, you should. Her data on the shape of the DNA molecule revealed its structure as a double helix. Watson and Crick get all the credit for identifying the double helix, but Franklin did much of the work. While researching the structure of DNA at King's College,

London, in the early 1950s, Franklin bounced X-rays off the molecule to produce incredibly sharp, detailed photos of it. Franklin's photos show a DNA molecule from the end, not the side, so it's difficult to envision the side view of the double helix you normally see. Yet Franklin knew she was looking at a helix.

Meanwhile, James Watson, a 23-year-old postdoctoral fellow at Cambridge, England, was working with a 38-year-old graduate student named Francis Crick. Together, they were building an enormous model of metal sticks and wooden balls, trying to figure out the structure of the same molecule Franklin had photographed.

Franklin was supposed to be collaborating with Maurice Wilkins, another scientist in her research group, but she and Wilkins despised each other (because of a switch in research projects in which Franklin was instructed to take over Wilkins's project without his knowledge). As their antagonism grew, so did Wilkins's friendship with Watson. What happened next is the stuff of science infamy. Just a few weeks before Franklin was ready to publish her findings, Wilkins showed Franklin's photographs of the DNA molecule to Watson — without her knowledge or permission! By giving Watson access to Franklin's data, Wilkins gave Watson and Crick the scoop on the competition.

Watson and Crick cracked the mystery of DNA structure using Chargaff's rules (see the section "Obeying Chargaff's rules" for details) and Franklin's measurements of the molecule. They deduced that the structure revealed by Franklin's photo, hastily drawn from memory by Watson, had to be a double helix, and Chargaff's rules pointed to bases in pairs. The rest of the structure came together like a big puzzle, and they rushed to publish their discovery in 1953. Franklin's paper, complete with the critical photos of the DNA molecule, was published in the same issue of the journal *Nature*.

In 1962, Watson, Crick, and Wilkins were honored with the Nobel Prize. Franklin wasn't properly credited for her part in their discovery but couldn't protest because she had died of ovarian cancer in 1957. It's quite possible that Franklin's cancer was the result of long-term exposure to X-rays during her scientific career. In a sense, Franklin sacrificed her life for science.

Chapter 7

Replication: Copying Your DNA

In This Chapter

▶ Uncovering the pattern for copying DNA

▶ Putting together a new DNA molecule

▶ Revealing how circular DNA molecules replicate

*E*verything in genetics relies on *replication* — the process of copying DNA accurately, quickly, and efficiently. Replication is part of reproduction (producing eggs and sperm), development (making all the cells needed by a growing embryo), and maintaining normal life (replacing skin, blood, and muscle cells).

Before meiosis can occur (see Chapter 2), the entire genome must be replicated so that a potential parent can make the eggs or sperm necessary for creating offspring. After fertilization occurs, the growing embryo must have the right genetic instructions in every cell to make all the tissues needed for life. As life outside the womb goes on, almost every cell in your body needs a copy of the entire genome to ensure that the genes that carry out the business of living are present and ready for action. For example, because you're constantly replacing your skin cells and white blood cells, your DNA is being replicated right now so that your cells have the genes they need to work properly.

This chapter explains all the details of the fantastic molecular photocopier that allows DNA — the stuff of life — to do its job. First, you tackle the basics of how DNA's structure provides a pattern for copying itself. Then, you find out about all the enzymes — those helpful protein workhorses — that do the labor of opening up the double-stranded DNA and assembling the building blocks of DNA into a new strand. Finally, you see how the copying process works, from beginning (origins) to ends (telomeres).

Unzipped: Creating the Pattern for More DNA

DNA is the ideal material for carrying genetic information because it

- Stores vast amounts of complex information *(genotype)* that can be "translated" into physical characteristics *(phenotype)*
- Can be copied quickly and accurately
- Is passed down from one generation to the next (in other words, it's *heritable*)

When James D. Watson and Francis Crick proposed the double helix as the structure of DNA (see Chapter 6 for coverage of DNA), they ended their 1953 paper with a pithy sentence about replication. That one little sentence paved the way for their next major publication, which hypothesized how replication may work. It's no accident that Watson and Crick won the Nobel Prize; their genius was uncanny and amazingly accurate. Without their discovery of the double helix, they never could've figured out replication, because the trick that DNA pulls off during replication depends entirely on how DNA is put together in the first place.

If you skipped Chapter 6, which focuses on how DNA is put together, you may want to skim over that material now. The main points about DNA you need to know to understand replication are:

- DNA is double-stranded.
- The nucleotide building blocks of DNA always match up in a complementary fashion — A (adenosine) with T (thymine) and C (cytosine) with G (guanine).
- DNA strands run antiparallel (that is, in opposite directions) to each other.

If you were to unzip a DNA molecule by breaking all the hydrogen bonds between the bases, you'd have two strands, and each would provide the pattern to create the other. During replication, special helper chemicals called *enzymes* bring matching (complementary) nucleotide building blocks to pair with the bases on each strand. The result is two exact copies built on the *templates* that the unzipped original strands provide.

Figure 7-1 shows how the original double-stranded DNA supplies a template to make copies of itself. This mode of replication is called *semiconservative*. No, this isn't how DNA may vote in the next election! In this case, semiconservative means that only half the molecule is "conserved," or left in its original state. (*Conservative,* in the genetic sense, means keeping something protected in its original state.)

Figure 7-1:
DNA provides its own pattern for copying itself using semiconservative replication.

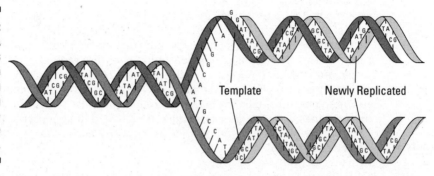

Template   Newly Replicated

At Columbia University in 1957, J. Herbert Taylor, Philip Woods, and Walter Hughes used the cell cycle to determine how DNA is copied (see Chapter 2 for a review of mitosis and the cell cycle). They came up with two possibilities: conservative or semiconservative replication.

Figure 7-2 shows how conservative replication may work. For both conservative and semiconservative replication, the original, double-stranded molecule comes apart and provides the template for building new strands. The result of semiconservative replication is two complete, double-stranded molecules, each composed of half "new" and half "old" DNA (which is what you see in Figure 7-1). Following conservative replication, the complete, double-stranded copies are composed of all "new" DNA, and the templates come back together to make one molecule composed of "old" DNA (as you can see in Figure 7-2).

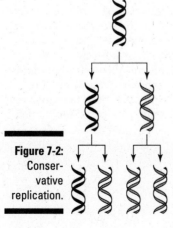

Figure 7-2:
Conservative replication.

To sort out replication, Taylor and his colleagues exposed the tips of a plant's roots to water that contained a radioactive chemical. This chemical was a form of the nucleotide building block *thymine,* which is found in DNA. Before cells in the root tips divided, their chromosomes incorporated the radioactive thymine as part of newly replicated DNA. In the first step of the experiment, Taylor and his team let the root tips grow for eight hours. That was just long enough for the DNA of the cells in the growing tips to replicate. The researchers collected some cells after this first step to see whether one or both sister chromatids of each chromosome were radioactive. Then, for the second step, they put the root tips in water with no radioactive chemical in it. After the cells started dividing, Taylor and his team examined the replicated chromosomes while they were in *metaphase* (when the replicated chromosomes, called *sister chromatids,* are all lined up together in the center of the cell, before they're pulled apart to opposite ends of the soon-to-divide cell; see Chapter 2).

The radioactivity allowed Taylor and his team to trace the fate of the template strands after replication was completed and determine whether the strands stayed together with their copies (semiconservative) or not (conservative). They examined the results of both steps of the experiment to ensure that their conclusions were accurate.

If replication was semiconservative, Taylor, Woods, and Hughes expected to find that one sister chromatid of the replicated chromosome would be radioactive and the other would be radiation-free — and that's what they got. Figure 7-3 shows how their results ended up as they did. The shaded chromosomes represent the ones containing the radioactive thymine. After one round of replication in the presence of the radioactive thymine (Step 1 in Figure 7-3), the entire chromosome appears radioactive.

If Taylor and his team could have seen the DNA molecules themselves (as you do figuratively here), they would have known that one strand of each double-stranded molecule contained radioactive thymine and the other did not (the radioactive strands are depicted with a thicker line). After one round of replication without access to the radioactive thymine (Step 2 in Figure 7-3), one sister chromatid was radioactive, and the other was not. That's because each strand from Step 1 provided a template for semiconservative replication: The radioactive strand provided one template, and the nonradioactive strand provided the other. After replication was completed, the templates remained paired with the new strands. This experiment showed conclusively that DNA replication is truly semiconservative — each replicated molecule of DNA is half "new" and half "old."

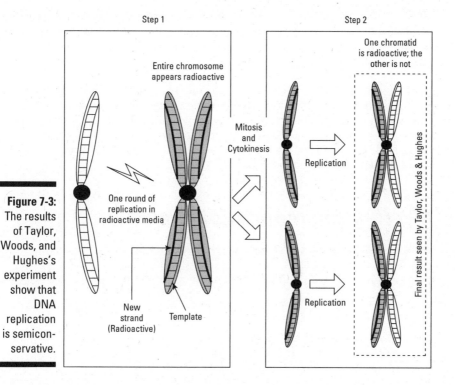

Step 1

Step 2

Entire chromosome
appears radioactive

One chromatid
is radioactive; the
other is not

Mitosis
and
Cytokinesis

Replication

One round of
replication in
radioactive media

New
strand
(Radioactive)

Template

Replication

Final result seen by Taylor, Woods & Hughes

Figure 7-3:
The results
of Taylor,
Woods, and
Hughes's
experiment
show that
DNA
replication
is semicon-
servative.

How DNA Copies Itself

Replication occurs during interphase of each cell cycle, just before prophase
in both mitosis and meiosis. If you skipped over Chapter 2, you may want
to take a quick glance at it to get an idea of when replication occurs with
respect to the life of a cell.

The process of replication follows a very specific order:

1. The helix is opened up to expose single strands of DNA.

2. Nucleotides are strung together to make new partner strands for the two
   original strands.

DNA replication was first studied in bacteria, which are *prokaryotic* (lacking
cell nuclei). All nonbacterial life forms (including humans) are *eukaryotes*
(composed of cells with nuclei). Prokaryotic and eukaryotic DNA replication
differ in a few ways. Basically, bacteria use slightly different versions of the

same enzymes that eukaryotic cells use, and most of those enzymes have similar names. If you understand prokaryotic replication, which I explain in this section, you have enough background to understand the details of eukaryotic replication, too.

Most eukaryotic DNA is linear, whereas most bacterial DNA (and your mitochondrial DNA) is circular. The shape of the chromosome (an endless loop versus a string) doesn't affect the process of replication at all. However, the shape means that circular DNAs have special problems to solve when replicating their hoop-shaped chromosomes. See the section "How Circular DNAs Replicate" later in this chapter to find out more.

Meeting the replication crew

For successful replication, several players must be present:

- ✔ **Template DNA,** a double-stranded molecule that provides a pattern to copy
- ✔ **Nucleotides,** the building blocks necessary to make new DNA
- ✔ **Enzymes and various proteins** that do the unzipping and assembly work of replication, called *DNA synthesis*

Template DNA

In addition to the material earlier in this chapter detailing how the template DNA is replicated semiconservatively (see "Unzipped: Creating the Pattern for More DNA"), it's vitally important for you to understand all the meanings of the term *template*.

- ✔ Every organism's DNA exists in the form of chromosomes. Therefore, the chromosomes undergoing replication and the template DNA uses during replication are one and the same.
- ✔ Both strands of each double-stranded original molecule are copied, and therefore, each of the two strands serves as a template (that is, a pattern) for replication.

The bases of the template DNA provide critical information needed for replication. Each new base of the newly replicated strand must be *complementary* (that is, an exact match; see Chapter 6 for more about the complementary nature of DNA) to the base opposite it on the template strand. Together, template and replicated DNA (like you see in Figure 7-1) make two identical copies of the original, double-stranded molecule.

Nucleotides

DNA is made up of thousands of nucleotides linked together in paired strands. (If you want more details about the chemical and physical constructions of DNA, flip to Chapter 6.) The nucleotide building blocks of DNA that come together during replication start in the form of *deoxyribonucleoside triphosphates,* or *dNTPs,* which are made up of

- A sugar (deoxyribose)
- One of four bases (adenine, guanine, thymine, or cytosine)
- Three phosphates

Figure 7-4 shows a dNTP being incorporated into a double-stranded DNA molecule. The dNTPs used in replication are very similar in chemical structure to the ones in double-stranded DNA (you can refer to Figure 6-3 in Chapter 6 to compare a nucleotide to the dNTP in Figure 7-4). The key difference is the number of phosphate groups — each dNTP has three phosphates, and each nucleotide has one.

Take a look at the blowup of the dNTP in Figure 7-4. The three phosphate groups (the "tri-" part of the name) are at the top end (usually referred to as the 5-prime, or 5') of the molecule. At the bottom left of the molecule, also known as the 3-prime (3') spot, is a little tail made of an oxygen atom attached to a hydrogen atom (collectively called an *OH group* or a *reactive group*). The oxygen atom in the OH tail is present to allow a nucleotide in an existing DNA strand to hook up with a dNTP; multiple connections like this one eventually produce the long chain of DNA. (For details on the numbered points of a molecule, such as 5' or 3', see Chapter 6.)

When DNA is being replicated, the OH tail on the 3' end of the last nucleotide in the chain reacts with the phosphates of a newly arrived dNTP (as shown in the right-hand part of Figure 7-4). Two of the dNTP's three phosphates get chopped off, and the remaining phosphate forms a phosphodiester bond with the previously incorporated nucleotide (see Chapter 6 for all the details about phosphodiester bonds). Hydrogen bonds form between the base of the template strand and the complementary base of the dNTP (see Chapter 6 for more on the bonds that form between bases). This reaction — losing two phosphates to form a phosphodiester bond and hydrogen bonding — converts the dNTP into a nucleotide. (The only real difference between dNTP and the nucleotide it becomes is the number of phosphates each carries.) Remember, the template DNA must be single-stranded for these reactions to occur (see "Splitting the helix" later in this chapter).

Each dNTP incorporated during replication must be complementary to the base it's hooked up with on the template strand.

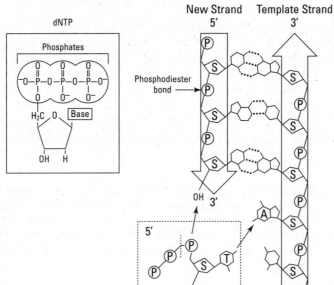

Figure 7-4:
Connecting
the chemical
building
blocks
(nucleotides
as dNTPs)
during DNA
synthesis.

A nucleotide is a deoxyribose sugar, a base, and a phosphate joined together as a unit. A nucleotide is a nucleotide regardless of whether it's part of a whole DNA molecule or not. A dNTP is also a nucleotide, just a special sort: a nucleotide triphosphate.

Enzymes

Replication can't occur without the help of a huge suite of enzymes. *Enzymes* are chemicals that cause reactions. Generally, enzymes come in two flavors: those that put things together and those that take things apart. Both types are used during replication.

Although you can't always tell the function of an enzyme (building or destroying) by its name, you can always identify enzymes because they end in *-ase*. The *-ase* suffix usually follows a reference to what the enzyme acts on. For example, the enzyme helicase acts on the helix of DNA to make it single-stranded (helix + ase = helicase).

So many enzymes are used in replication that it's hard to keep up with them all. However, the main players and their roles are:

✓ **Helicase:** Opens up the double helix

✓ **Gyrase:** Prevents the helix from forming knots

✔ **Primase:** Lays down a short piece of RNA (a primer) to get replication started (see Chapter 8 for more on RNA)

✔ **DNA polymerase:** Adds dNTPs to build the new strand of DNA

✔ **Ligase:** Seals the gaps between newly replicated pieces of DNA

✔ **Telomerase:** Replicates the ends of chromosomes (the telomeres) — a very special job

Prokaryotes have 5 forms of DNA polymerase, and eukaryotes have at least 13 forms. In prokaryotes, DNA polymerase III is the enzyme that performs replication. DNA polymerase I removes RNA primers and replaces them with DNA. DNA polymerases II, IV, and V all work to repair damaged DNA and carry out proofreading activities. Eukaryotes use a whole different set of DNA polymerases. (For more details on eukaryotic DNA replication, see the section "Replication in Eukaryotes" later in the chapter.)

Splitting the helix

DNA replication starts at very specific spots, called *origins,* along the double-stranded template molecule. Bacterial chromosomes are so short (only about 4 million base pairs; see Chapter 11) that only one origin for replication is needed. Copying bigger genomes would take far too long if each chromosome had only one origin, so to make the process of copying very rapid, human chromosomes each have thousands of origins. (See the section "Replication in Eukaryotes" later in this chapter for more details on how human DNA is replicated.)

Special proteins called *initiators* move along the double-stranded template DNA until they encounter a group of bases that are in a specific order. These bases represent the origin for replication; think of them as a road sign with the message: "Start replication here." The initiator proteins latch onto the template at the origin by looping the helix around themselves like looping a string around your finger. The initiator proteins then make a very small opening in the double helix.

Helicase (the enzyme that opens up the double helix) finds this opening and starts breaking the hydrogen bonds between the complementary template strands to expose a few hundred bases and split the helix open even wider. DNA has such a strong tendency to form double-strands that if another protein didn't come along to hold the single strands exposed by helicase apart, they'd snap right back together again. These proteins, called *single-stranded-binding* (SSB) proteins, prop the two strands apart so replication can occur. Figure 7-5 shows the whole process of replication. For now, focus on the part that shows how helicase breaks the strands apart as it moves along the double helix and how the strands are kept separated and untwisted.

If you've had any experience with yarn or fishing line, you know that if string gets twisted together and you try to pull the strands apart, a knot forms. This same problem occurs when opening up the double helix of DNA. When helicase starts pulling the two strands apart, the opening of the helix sends extra turns along the intact helix. To prevent DNA from ending up a knotty mess, an enzyme called *gyrase* comes along to relieve the tension. Exactly how gyrase does this is unclear, but some researchers think that gyrase actually snips the DNA apart temporarily to let the twisted parts relax and then seals the molecule back together again.

Priming the pump

When helicase opens up the molecule, a Y forms at the opening. This Y is called a *replication fork*. You can see a replication fork in Figure 7-5, where the helicase has split the DNA helix apart. For every opening in the double-stranded molecule, two forks form on opposite sides of the opening. DNA replication is very particular in that it can only proceed in one direction: 5-prime to 3-prime ($5' \rightarrow 3'$). In Figure 7-5, the top strand runs $3' \rightarrow 5'$ from left to right, and the bottom strand runs $5' \rightarrow 3'$ (that is, the template strands are *antiparallel*; see Chapter 6 for more about the importance of the antiparallel arrangement of DNA strands).

Replication must proceed antiparallel to the template, running 5' to 3'. Therefore, replication on the top strand runs right to left; on the bottom strand, replication runs left to right.

After helicase splits the molecule open (as I explain in the preceding section), two naked strands of template DNA are left. Replication can't start on the naked template strands because it hasn't started yet. (That sounds a bit like Yogi Berra saying "It ain't over 'til it's over," doesn't it?) All funny business aside, nucleotides can only form chains if a nucleotide is already present with a free reactive tail on which to attach the incoming dNTP. DNA solves the problem of starting replication by inserting *primers*, little complementary starter strands made of RNA (refer to Figure 7-5).

Primase, the enzyme that manufactures the RNA primers for replication, lays down primers at each replication fork so that DNA synthesis can proceed from $5' \rightarrow 3'$ on both strands. The RNA primers made by primase are only about 10 or 12 nucleotides long. They're complementary to the single strands of DNA and end with the same sort of OH tail found on a nucleotide of DNA. (To find out more about RNA, flip to Chapter 8.) DNA uses the primers' free OH tails to add nucleotides in the form of dNTPs (see "Nucleotides" earlier in this chapter); the primers are later snipped out and replaced with DNA (see "Joining all the pieces" later in this chapter).

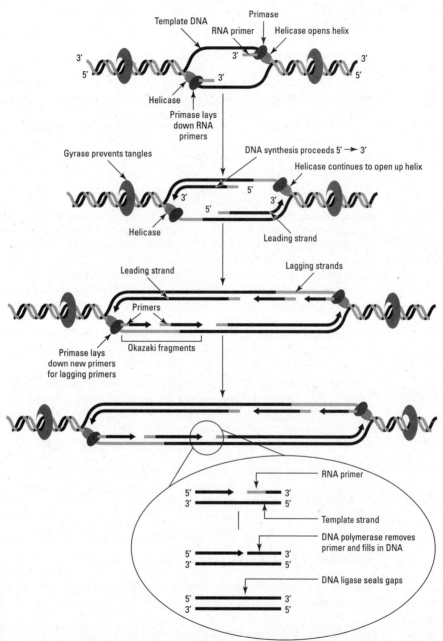

Figure 7-5:
The process of replication.

Leading and lagging

As soon as the primers are in place, actual replication can get underway. *DNA polymerase* is the enzyme that does all the work of replication. At the OH tail of each primer, DNA polymerase tacks on dNTPs by snipping off two phosphates and forming phosphodiester bonds (see Chapter 6). Meanwhile, helicase opens up the helix ahead of the growing chain to expose more template strand. From Figure 7-5, it's easy to see that replication can just zoom along this way — but only on one strand (in this case, the top strand in Figure 7-5). The replicated strands keep growing continuously 5' → 3' as helicase makes the template available.

At the same time, on the opposite strand, new primers have to be added to take advantage of the newly available template. The new primers are necessary because a naked strand (the bottom one in Figure 7-5) lacking the necessary free nucleotide for chain-building is created by the ongoing splitting of the helix.

Thus, the interaction of opening the helix and synthesizing DNA 5' → 3' on one strand while laying down new primers on the other leads to the formation of *leading* and *lagging strands*.

- ✔ **Leading strands:** The strands formed in one bout of uninterrupted DNA synthesis (you can see a leading strand in Figure 7-6). Leading strands follow the lead, so to speak, of helicase.

- ✔ **Lagging strands:** The strands that are begun over and over as new primers are laid down. Synthesis of the lagging strands stops when they reach the 5' end of a primer elsewhere on the strand. Lagging strands "lag behind" leading strands in the sense of frequent starting and stopping versus continuous replication. (Replication happens so rapidly that there's no difference in the amount of time it takes to replicate leading and lagging strands.) The short pieces of DNA formed by lagging DNA synthesis have a special name: *Okazaki fragments,* named for the scientist, Reiji Okazaki, who discovered them.

Joining all the pieces

After the template strands are replicated, the newly synthesized strands have to be modified to be complete and whole:

- ✔ The RNA primers must be removed and replaced with DNA.
- ✔ The Okazaki fragments formed by lagging DNA synthesis must be joined together.

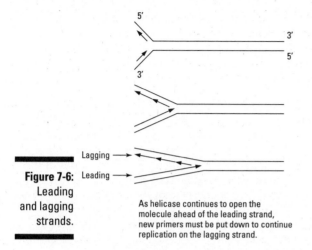

Figure 7-6:
Leading
and lagging
strands.

As helicase continues to open the
molecule ahead of the leading strand,
new primers must be put down to continue
replication on the lagging strand.

A special kind of DNA polymerase moves along the newly synthesized
strands seeking out the RNA primers. When DNA polymerase encounters the
short bits of RNA, it snips them out and replaces them with DNA. (Refer to
Figure 7-5 for an illustration of this process.) The snipping out and replacing
of RNA primers proceeds in the usual 5' → 3' direction of replication and follows the same procedures as normal DNA synthesis (adding dNTPs and forming phosphodiester bonds).

After the primers are removed and replaced, one phosphodiester bond is
missing between the Okazaki fragments. *Ligase* is the enzyme that seals these
little gaps (*ligate* means to join things together). Ligase has the special ability
to form phosphodiester bonds without adding a new nucleotide.

Proofreading replication

Despite its complexity, replication is unbelievably fast. In humans, replication speeds along at about 2,000 bases a minute. Bacterial replication is even
faster at about 1,000 bases per *second!* Working at that speed, it's really no
surprise that DNA polymerase makes mistakes — about one in every 100,000
bases is incorrect. Fortunately, DNA polymerase can use the backspace key!

DNA polymerase constantly checks its work though a process called
proofreading — the same way I proofread my work as I wrote this book.
DNA polymerase looks over its shoulder, so to speak, and keeps track of
how well the newly added bases fit with the template strand. If an incorrect
base is added, DNA polymerase backs up and cuts the incorrect base out.
The snipping process is called *exonuclease activity,* and the correction

process requires DNA polymerase to move 3' → 5' instead of the usual 5' → 3' direction. DNA proofreading eliminates most of the mistakes made by DNA polymerase, and the result is nearly error-free DNA synthesis. Generally, replication (after proofreading) has an astonishingly low error rate of one in 10 million base pairs.

If DNA polymerase misses an incorrect base, special enzymes come along after replication is complete to carry out another process, called *mismatch repair* (much like my editors checked my proofreading). The mismatch repair enzymes detect the bulges that occur along the helix when noncomplementary bases are paired up, and the enzymes snip the incorrect base out of the newly synthesized strand. These enzymes replace the incorrect base with the correct one and, like ligase, seal up the gaps to finish the repair job.

Replication is a complicated process that uses a dizzying array of enzymes. The key points to remember are:

✔ Replication always starts at an origin.

✔ Replication can only occur when template DNA is single-stranded.

✔ RNA primers must be put down before replication can proceed.

✔ Replication always moves 5' → 3'.

✔ Newly synthesized strands are exact complementary matches to template ("old") strands.

Replication in Eukaryotes

Although replication in prokaryotes and eukaryotes is very similar, you need to know about four differences:

✔ For each of their chromosomes, eukaryotes have many, many origins for replication. Prokaryotes generally have one origin per circular chromosome.

✔ The enzymes that prokaryotes and eukaryotes use for replication are similar but not identical. Compared to prokaryotes, eukaryotes have many more DNA polymerases, and these DNA polymerases carry out other functions besides replication.

✔ Linear chromosomes, found in eukaryotes, require special enzymes to replicate the *telomeres* — the ends of chromosomes.

✔ Eukaryotic chromosomes are tightly wound around special proteins in order to package large amounts of DNA into very small cell nuclei.

Pulling up short: Telomeres

When linear chromosomes replicate, the ends of the chromosomes, called *telomeres,* present special challenges. These challenges are handled in different ways depending on what kind of cell division is taking place (that is, mitosis versus meiosis).

At the completion of replication for cells in mitosis, a short part of the telomere tips is left single-stranded and unreplicated. A special enzyme comes along and snips off this unreplicated part of the telomere. Losing this bit of DNA at the end of the chromosome isn't as big a deal as it may seem, because telomeres, in addition to being the ends of chromosomes, are long strings of *junk DNA*. Junk DNA doesn't contain genes but may have other important functions (see Chapter 11 for the details).

For telomeres, being junk DNA is good because when telomeres get snipped off, the chromosomes aren't damaged too much and the genes still work just fine — up to a point. After many rounds of replication, all the junk DNA at the ends of the chromosomes is snipped off (essentially, the chromosomes run out of junk DNA), and actual genes themselves are affected. Therefore, when the chromosomes of a mitotic cell (like a skin cell, for example) get too short, the cell dies through a process called *apoptosis.* (I cover apoptosis in detail in Chapter 14.) Paradoxically, cell death through apoptosis is a good thing because it protects you from the ravages of mutations, which can cause cancer.

If the cell is being divided as part of meiosis, telomere snipping is not okay. The telomeres must be replicated completely so that perfectly complete, full-size chromosomes are passed on to offspring. An enzyme called *telomerase* takes care of replicating the ends of the chromosomes. Figure 7-7 gives you an idea of how telomerase replicates telomeres. Primase lays down a primer at the very tip of the chromosome as part of the normal replication process. DNA synthesis proceeds from 5' → 3' as usual, and then, a DNA polymerase comes along and snips out the RNA primer from 5' → 3'. Without telomerase, the process stops, leaving a tail of unreplicated, single-stranded DNA flapping around (this is what happens during mitosis).

Telomerase easily detects the unreplicated telomere because telomeres have long sections of guanines, or Gs. Telomerase contains a section of cytosine-rich RNA, allowing the enzyme to bind to the unreplicated, guanine-rich telomere. Telomerase then uses its own RNA to extend the unreplicated DNA template by about 15 nucleotides. Scientists suspect that the single-stranded template then folds back on itself to provide a free OH tail to replicate the rest of the telomere in the absence of a primer (see "Priming the pump" earlier in this chapter).

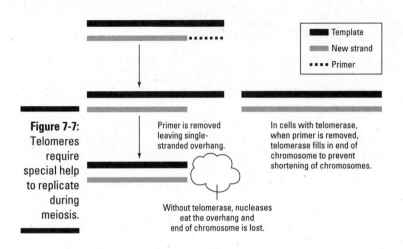

■■■	Template
▬▬▬	New strand
•••••	Primer

Figure 7-7:
Telomeres require special help to replicate during meiosis.

Primer is removed leaving single-stranded overhang.

In cells with telomerase, when primer is removed, telomerase fills in end of chromosome to prevent shortening of chromosomes.

Without telomerase, nucleases eat the overhang and end of chromosome is lost.

Finishing the job

Your DNA (and that of all eukaryotes) is tightly wound around special struc-tures called *nucleosomes* (not to be confused with nucleotides) so that the enormous molecule fits neatly into the cell nucleus. (See Chapter 6 for the details on just how big a molecule of DNA really is.) Like replication, packag-ing DNA is a very rapid process. It happens so quickly that scientists aren't exactly sure how DNA gets unwrapped from the nucleosomes to replicate and then gets wrapped around the nucleosomes again.

In the packaging stage, DNA is normally twisted tightly around hundreds of thousands of nucleosomes, much like string wrapped around beads. The whole "necklace" gets wound very tightly around itself in a process called *supercoil-ing*. Supercoiling is what allows the 3.5 billion base pairs of DNA that make up your 46 chromosomes to fit inside the microscopic nuclei of your cells. Altogether, about 150 base pairs of DNA are wrapped around each nucleosome and secured in place with a little protein called a *histone*. In Figure 7-8, you can see the nucleosomes, histones, and supercoiled "necklace."

DNA is packaged in this manner both before and after replication. Because only 30 or 40 base pairs of DNA are exposed between nucleosomes, the DNA must be removed from the nucleosomes in order to replicate. If it isn't removed from the nuclesomes, the enzymes used in replication aren't able to access the entire molecule.

As helicase opens up the DNA molecule during replication, an unidentified enzyme strips off the nucleosome beads at the same time. As soon as the DNA is replicated, the DNA (both old and new) is immediately wrapped around waiting nucleosomes. Studies show that the old nucleosomes (from before replication) are reused along with newly assembled nucleosomes to package the freshly replicated DNA molecule.

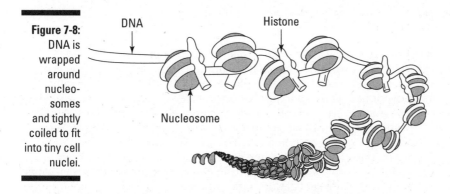

DNA

Histone

Nucleosome

How Circular DNAs Replicate

Circular DNAs are replicated in three different ways, as shown in Figure
7-9. Different organisms take different approaches to solve the problem of
replicating hoop-shaped chromosomes. Theta replication is used by most
bacteria, including _E. coli._ Viruses use rolling circle replication to rapidly
manufacture vast numbers of copies of their genomes. Finally, human mitochondrial DNA and the chloroplast DNA of plants both use D-loop replication.

Theta

Theta replication refers to the shape the chromosome takes on during the
replication process. After the helix splits apart, a bubble forms, giving the
chromosome a shape reminiscent of the Greek letter theta (Θ; see Figure 7-9).
Bacterial chromosomes have only one origin of replication (see "Splitting the
helix"), so after helicase opens the double helix, replication proceeds in both
directions simultaneously, rapidly copying the entire molecule. As I describe
in the section "Leading and lagging," leading and lagging strands form, and
ligase seals the gaps in the newly synthesized DNA to complete the strands.
Ultimately, theta replication produces two intact, double-stranded molecules.

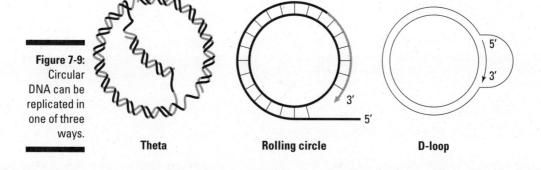

Figure 7-9:
Circular
DNA can be
replicated in
one of three
ways.

Theta   **Rolling circle**   **D-loop**

Rolling circle

Rolling circle replication creates an odd situation. No primer is needed because the double-stranded template is broken at the origin to provide a free OH tail to start replication. As replication proceeds, the inner strand is copied continuously as a leading strand (refer to Figure 7-9). Meanwhile, the broken strand is stripped off. As soon as enough of the broken strand is freed, a primer is laid down so replication can occur as the broken strand is stripped away from its complement. Thus, rolling circle replication is continuous on one strand and lagging on the other. As soon as replication is completed for one copy of the genome, the new copies are used as templates for additional rounds of replication. Viral genomes are often very small (only a few thousand base pairs), so rolling circle replication is an extremely rapid process that produces hundreds of thousands of copies of viral DNA in only a few minutes.

D-loop

Like rolling circle replication, *D-loop replication* creates a displaced, single strand (refer to Figure 7-9). Helicase opens the double-stranded molecule, and an RNA primer is laid down, displacing one strand. Replication then proceeds around the circle, pushing the displaced strand off as it goes. The intact, single strand is released and used as a template to synthesize a complementary strand.

Chapter 8

Sequencing Your DNA

*I*magine owning a library of 22,000 books. I don't mean just any books; this collection contains unimaginable knowledge, such as solutions to diseases that have plagued humankind for centuries, basic building instructions for just about every creature on earth, and even the explanation of how thoughts are formed inside your brain. This fabulous library has only one problem — it's written in a mysterious language, a code made up of only four letters that are repeated in arcane patterns. The very secrets of life on earth have been contained within this library since the dawn of time, but no one could read the books — until now.

The 22,000 books are the genes that carry the information that make you. The library storing these books is the human genome. Sequencing *genomes* (that is, all the DNA in one set of chromosomes of an organism), both human genomes and those of other organisms, means discovering the order of the four bases (C, G, A, and T) that make up DNA. The order of the bases in DNA is incredibly important because it's the key to DNA's language, and understanding the language is the first step in reading the books of the library. Most of your genes are identical to those in other species, so sequencing the DNA of other organisms, such as fruit flies, roundworms, chickens, and even yeast, supplies scientists with a lot of information about the human genome and how human genes function.

Trying on a Few Genomes

Humans are incredibly complex organisms, but when it comes to genetics, they're not at the top of the heap. Many complex organisms have vastly larger genomes than humans do. Genomes are usually measured in the number of base pairs they contain (flip to Chapter 6 for more about how DNA is put together in base pairs). Table 8-1 lists the genome sizes and estimated number of genes for various organisms (for some genomes, like grasshoppers, the numbers of genes are still unknown). Human genome size runs a distant fifth behind salamanders, amoebas, and grasshoppers. It's humbling, but true — a single-celled amoeba has a gigantic genome of over 670 billion base pairs. If genome size and complexity were related (and they obviously aren't), you'd expect the amoeba to have a small genome compared to more complex organisms. On the flip side, it doesn't take a lot of DNA to have a big impact on the world. For example, the HIV virus, which causes AIDS, is a mere 9,700 bases long and is responsible for the deaths of over 25 million people worldwide. With only nine genes, HIV isn't very complex, but it's still very dangerous.

Even organisms that are very similar have vastly different genome sizes. Fruit flies have roughly 180 million base pairs of DNA. Compare that to the grasshopper genome, which weighs in at a whopping 180 *billion* base pairs. But fruit flies and grasshoppers aren't *that* different. So if it isn't organism complexity, what causes the differences in genome size among organisms?

Table 8-1	Genome Sizes of Various Organisms	
Species	*Number of Base Pairs*	*Number of Genes*
HIV virus	9,700	9
E. coli	4,600,000	3,200
Yeast	12,000,000	6,532
Flu bacteria	19,000,000	1,700
Roundworm	103,000,000	20,158
Mustard weed	120,000,000	27,379
Fruit fly	180,000,000	14,422
Chicken	1,000,000,000	15,926
Mouse	3,400,000,000	22,974
Corn	2,500,000,000	50,000–60,000
Human	3,000,000,000	22,258
Grasshopper	180,000,000,000	unknown
Amoeba dubia	670,000,000,000	unknown
Salamander	765,000,000,000	unknown

Part of what accounts for the variation in genome size from one organism to the next is number of chromosomes. Particularly in plants, the number of chromosome sets (called *ploidy;* see Chapter 15) explains why some plant species have very large genome sizes. For example, wheat is *hexaploid* (six copies of each chromosome) and has a gigantic genome of 16 billion base pairs. Rice, on the other hand, is *diploid* (two copies of each chromosome) and has a mere 430 million base pairs.

Chromosome number doesn't tell the whole story, however. The number of genes within a genome doesn't reveal how big the genome is. Arguably, mice are somewhat more complex than corn, but they have at least 27,000 fewer genes! On top of that, the mouse genome is larger than the corn genome by about a million base pairs. What the human genome has that the mustard weed genome may lack is lots of repetition.

DNA sequences fall into two major categories:

- ✔ Unique sequences found in genes (I cover genes in Chapter 11)
- ✔ Repetitive sequences that make up noncoding DNA

The presence of repetitive sequences of DNA in some organisms seems to best explain genome size — that is, large genomes have many repeated sequences that smaller genomes lack. Repetitive sequences vary from 150 to 300 base pairs in length and are repeated thousands and thousands of times. These big chunks of sequences don't code for proteins, though. Because, at least initially, all this repetitive DNA didn't seem to do anything, it was dubbed *junk DNA*.

Junk DNA has suffered a bum rap. For years, it was touted as a genetic loser, just along for the ride, doing nothing except getting passed on from one generation to the next. But no more. At long last, so-called junk DNA is getting proper respect. Scientists realized quite some time ago that a lot of DNA besides genes gets transcribed into RNA (see Chapter 9 for more on the transcription process). But after being transcribed, this noncoding "junk" didn't appear to be translated into protein (see Chapter 10 for more on the translation process). New evidence suggests that repeated sequences control transcription. A recent study identified 200,000 transcription start sites within repetitive DNA in the human and mouse genomes, suggesting that "junk" DNA may turn out to be the most important part of the genome, controlling everything from how organisms develop as embryos to what color your eyes are.

Sequencing Your Way to the Human Genome

One of the ways scientists figure out what functions various kinds of sequences carry out is by comparing genomes of different organisms. To make these comparisons, the projects I describe in this section use the methods I explain in the section "Sequencing: Reading the Language of DNA" later in this chapter. The results of these comparisons tell us a lot about ourselves and the world around us.

The DNA of all organisms holds a vast amount of information. Amazingly, most cell functions work the same, regardless of which animal the cell comes from. Yeast, elephants, and humans all replicate DNA in the same way, using almost identical genes. Because nature uses the same genetic machinery over and over, finding out about the DNA sequences in other organisms tells us a lot about the human genome (and it's far easier to experiment with yeast and roundworms than with humans). Table 8-2 is a timeline of the major milestones of DNA sequencing projects so far. In this section, you find out about several of these projects, including the granddaddy of them all, the Human Genome Project.

Table 8-2	Major Milestones in DNA Sequencing
Year	*Event*
1985	Human Genome Project is proposed.
1990	Human Genome Project officially begins.
1992	First map of all genes in the entire human genome is published.
1995	First sequence of an entire living organism — *Haemophilus influenzae,* a flu bacteria — is completed.
1997	Genome of *Escherichia coli,* the most common intestinal bacteria, is completed.
1999	First human chromosome, chromosome 22, is completely sequenced. Human Genome Project passes the 1 billion base pairs milestone.
2000	Fruit fly genome is completed. First entire plant genome — *Arabidopsis thaliana,* the common mustard plant — is sequenced.
2001	First working "draft" of the entire human genome is published.
2002	Mouse genome is completed.
2004	Chicken genome is completed, as is the *euchromatin* (gene-containing) sequence of the human genome.
2006	Cancer Genome Atlas project launched.
2008	First high-resolution map of genetic variation among humans is published.

The yeast genome

Brewer's yeast (scientific name *Saccharomyces cerevisiae*) was the first eukaryotic genome to be fully sequenced. (*Eukaryotes* have cells with nuclei; see Chapter 2.) Yeast has an established track record as one of the most useful organisms known to humankind. It's responsible for making bread rise and for the fermentation that results in beer and wine. It's also a favorite organism for genetic study. Much of what we know about the eukaryotic cell cycle (see Chapter 2) came from yeast research. Yeast has provided information about how genes are inherited together (called linkage; see Chapter 4) and how genes are turned on and off (see Chapter 10). Because many human genes have yeast counterparts, yeast is extremely valuable for finding out how our own genes work.

Yeast has roughly 6,000 genes and 16 chromosomes. Altogether, about 70 percent of the yeast genome consists of actual genes. Yeast genes work in neighborhoods to carry out their functions; genes that are physically close together on chromosomes are more likely to work together than those that are far apart. The discovery of gene networks in yeast may help researchers better understand what causes complex diseases such as Alzheimer's disease, diabetes, and lupus in humans. Disorders such as these aren't typically inherited in simple Mendelian fashion (see Chapter 3) and are likely to be controlled by many genes working together.

The sequencing of the yeast genome was quite a feat. Over 600 researchers in 100 laboratories across the world participated in the project. The technology used at the time was much slower than what's available to researchers now (see the sidebar "Open access and the Human Genome Project" for details). Despite the technological disadvantage, the sequence that this phenomenal team of scientists produced was extremely accurate — especially when compared to the human genome (see "The Human Genome Project" section later in this chapter).

The elegant roundworm genome

The genome of the lowly roundworm, more properly referred to by its full name *Caenorhabditis elegans,* was the first genome of a multicellular organism to be fully sequenced. Weighing in at roughly 97 million base pairs, the roundworm boasts nearly 20,000 genes — only a few thousand fewer than the human genome — on just six chromosomes. Like humans, roundworms have lots of junk DNA; only 25 percent of the roundworm genome is made up of genes.

Roundworms are a fabulous species to study because they reproduce sexually and have organ systems, such as digestive and nervous systems, similar to those in much more complex organisms. Additionally, roundworms have a sense of taste, can detect odors, and react to light and temperature, so they're ideal for studying all sorts of processes, including behavior. Full-grown roundworms have exactly 959 cells and are transparent, so figuring out how their cells work was relatively easy. Scientists determined the exact function of each of the 959 roundworm cells! Although roundworms live in soil, these microscopic organisms have contributed to our understanding of many human diseases.

One of the ways to discover what a gene does is to stop it from functioning and observe the effect. In 2003, a group of researchers fed roundworms a particular kind of RNA that temporarily puts gene function on hold (see Chapter 10 for how this effect on gene function works). By briefly turning genes off, the scientists were able to determine the functions of roughly 16,000 of the round-worms' genes. Another study using the same technique identified how fat storage and obesity are controlled in roundworms. Given that an amazing 70 percent of proteins that humans produce have roundworm counterparts, these gene function studies have obvious implications for human medicine.

The chicken genome

Chickens don't get enough respect. The study of chicken biology has revealed much about how organisms develop from embryos to adults. For example, a study of how a chicken's wings and legs are formed in the egg greatly enhanced a study of human limb formation. Chickens have contributed to our understanding of diseases such as muscular dystrophy and epilepsy, and chicken eggs are the principal ingredient used to produce vaccinations to fight human disease epidemics. So when the chicken genome was sequenced in 2004, there should have been a lot of crowing about the underappreciated chicken.

The chicken genome is really different from mouse and human genomes. It's much smaller (about a third as big as the human genome), with fewer chromosomes (39 compared to our 46) and a similar number of genes (23,000 or so). Roughly 60 percent of chicken genes have human counterparts. Unlike mammals, some chicken chromosomes are tiny (only about 5 million base pairs). These micro-chromosomes are unique because they have a very high content of guanine and cytosine (see Chapter 6 for more about the bases that make up DNA) and very few repetitive sequences.

Not surprisingly, chickens have lots of genes that code for *keratin* — the stuff that makes their feathers (and your hair). The big surprise regarding the completed chicken genome was that chickens have lots of genes for sense of smell. Until recently, scientists thought that most birds have a really poor sense of smell. Now, they realize that sense of taste is what birds lack. The chicken genome also revealed that a particular gene previously known only to exist in humans is also present in chickens. This gene, called *interleukin 26,* is important in immune responses and may allow researchers to better understand how to fight disease. One disease they're particularly interested in is avian flu which is often deadly to humans but doesn't make birds sick. Ultimately, comparing the chicken and human genomes may allow scientists to understand how and why diseases like the "bird flu" move so easily between chickens and humans.

The Human Genome Project

In 2001, the triumphant publication of the human genome sequence was heralded as one of the great feats of modern science. The sequence was considered a draft, and indeed, it was a really *rough* draft. The 2001 sequence was woefully incomplete (it represented only about 60 percent of the total human genome) and was full of errors that limited its utility. In 2004, the *euchromatic* (or gene-containing) sequence had only a few gaps, and most of the errors had been corrected. By 2008, new technologies allowed comparisons between individual humans, laying the foundation for a better understanding of how genes vary to create the endless phenotypes you see around you.

The Human Genome Project (HGP) is akin to some of the greatest adventures of all time — it's not unlike putting a person on the moon. However, unlike the great technological achievements of space exploration, which cost tens of billions of dollars and require technology that becomes obsolete or wears out, the HGP carries a mere $3 billion price tag and has unlimited utility. When first proposed in 1985, the HGP was considered completely impossible. At that time, sequencing technology was slow, requiring several days to generate only a few hundred base pairs of data (see the sidebar "Open access and the Human Genome Project" to find out how this process was sped up). James Watson, one of the discoverers of DNA structure way back in the 1950s (see Chapter 6), was one of the first to push the project (in 1988) from idea to reality during his tenure as director of the National Institutes of Health. When the project got off the ground in 1990, a global team of scientists from 20 institutions participated. (The 2001 human genome sequence paper had a staggering 273 authors.)

The enormous benefits of the HGP remain underappreciated. Most genetic applications wouldn't exist without the HGP. Here are just a few:

- Development of bioinformatics, an entirely new field focused on advancing technological capability to generate genetic data, catalog results, and compare genomes (flip to Chapter 23 for more).

- Development of drugs and gene therapy (see Chapter 16).

- Diagnosis and treatment of genetic disorders (which I cover in Chapter 12).

- Forensics applications, such as identification of criminals and determination of identity after mass disasters (flip to Chapter 18).

- Generation of thousands of jobs and economic benefits of over $25 billion in one year alone (2001).

- Identification of bacteria and viruses to allow for targeted treatment of disease. Some antibiotics, for example, target some strains of bacteria better than others. Genetic identification of bacteria is quick and inexpensive, allowing physicians to rapidly identify and prescribe the right antibiotic.

- Knowledge of which genes control what functions and how those genes are turned on and off (see Chapter 11).

- Understanding of the causes of cancer (which I cover in Chapter 14).

Listing and explaining all the HGP's discoveries would fill this book and then some. As you can see in Table 8-2, all other genome projects — mouse, fruit fly, yeast, roundworm, mustard weed, and so on — were started as a result of the HGP.

As the HGP progressed, the gene count in the human genome steadily declined. Originally, researchers thought that humans had as many as 100,000 genes. But as new and more accurate information has become available over the years, they've determined that the human genome has only about 22,000 genes. Genes are often relatively small, from a base-pair standpoint (roughly 3,000 base pairs), meaning that less than 2 percent of your DNA actually codes for some protein. The number of genes on different chromosomes varies enormously, from nearly 3,000 genes on chromosome 1 (the largest) to 231 genes on the Y chromosome (the smallest).

The Human Genome Project has revealed the surprisingly dynamic and still changing nature of the human genome. One of the surprising discoveries of the HGP is that the human genome is still "growing." Genes get duplicated and then gain new functions, a process that has produced as many as 1,100 new genes. Likewise, genes lose function and "die." Thanks to this death process, 37 genes in the human genome that were once functional now exist as *pseudogenes,* which have the sequence structure of normal genes but no longer code for proteins (see Chapter 11 for more about genes).

Of the human genes that researchers have identified, they only understand what about half of them do. Comparisons with genomes of other organisms help identify what genes do, because most of the proteins that human genes produce have counterparts in other organisms. Thus, humans share many genes in common with even the simplest organisms, such as bacteria and worms. Over 99 percent of your DNA is identical to that of any other human on earth, and as much as 98 percent of your DNA is identical to the sequences found in the mouse genome. Perhaps the greatest take-home message of the HGP is how alike all life on earth really is.

Sequencing: Reading the Language of DNA

The chemical nature of DNA (which I cover in Chapter 6) and the replication process (which you can discover in Chapter 7) are essential to DNA sequencing. DNA sequencing also makes use of a reaction that's similar to the polymerase chain reaction (PCR) used in forensics; if you want more details about PCR, check out Chapter 18.

Identifying the players in DNA sequencing

New technologies are rapidly changing the way DNA sequencing is done (see the sidebar "Open access and the Human Genome Project" for more info). The old tried-and-true approach I describe here, called the Sanger method (after inventor Frederick Sanger), still provides the basis for many of the new methods.

The key ingredients for DNA sequencing are:

✔ **DNA:** From a single individual of the organism to be sequenced.

✔ **Primers:** Several thousand copies of short sequences of DNA that are complementary to the part of the DNA to be sequenced.

✔ **dNTPs:** Many As, Gs, Cs, and Ts, put together with sugars and phosphates as *nucleotides,* the normal building blocks of DNA.

✔ **ddNTPs:** Many As, Gs, Cs, and Ts as nucleotides that each lack an oxygen atom at the 3' spot.

✔ **Taq polymerase:** The enzyme that puts the DNA molecule together (see Chapter 18 for more details on Taq).

The use of ddNTPs is the key to how sequencing works. Take a careful look at Figure 8-1. On the left is a generic dNTP, the basic building block of DNA used during replication (if you don't remember all the details, flip to Chapter 6 for more on dNTPs). The molecule on the right is ddNTP *(dideoxyribonucleoside triphosphate).* The ddNTP is identical to the dNTP in every way except that it has no oxygen atom at the 3' spot. No oxygen means no reaction, because the phosphate group of the next nucleotide can't form a phosphodiester bond (see Chapter 6) without that extra oxygen atom to aid the reaction. The next nucleotide can't hook up to ddNTP at the end of the chain, and the replication process stops. So how does *stopping* the reaction help the sequencing process? The idea is to create thousands of short pieces of DNA that give the identity of each and every base along the sequence.

The result of a typical sequencing reaction is a thousand fragments representing a thousand bases of the template strand. The shortest fragment is made up of a primer and one ddNTP representing the complement of the first base of the template. The next shortest fragment is made up of the primer, one nucleotide (from a dNTP), and a ddNTP — and so on, with the largest fragment being a thousand bases long.

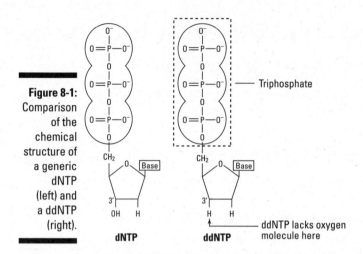

Figure 8-1:
Comparison
of the
chemical
structure of
a generic
dNTP
(left) and
a ddNTP
(right).

Finding the message in sequencing results

To see the results of the sequencing reaction, scientists put the DNA fragments through a process called *electrophoresis*. Electrophoresis is the
movement of charged particles (in this case, DNA) under the influence of
electricity. The purpose of electrophoresis is to sort the fragments of DNA by
size, from smallest to largest. The smallest fragment gives the first base in the
sequence, the second-smallest fragment gives the second base, and so on,
until the largest fragment gives the last base in the sequence. This arrangement of fragments allows researchers to read the sequence in its proper
order.

A computer-driven machine called a *sequencer* uses a laser to see the colored
dyes of the ddNTPs at the end of each fragment. The laser shines into the gel
and reads the color of each fragment as it passes by. Fragments pass the laser
in order of size, from smallest to largest. Each dye color signals a different
letter: As show up green, Ts are red, Cs are blue, and Gs are yellow. The computer automatically translates the colors into letters and stores all the information for later analysis.

The resulting picture is a series of peaks, like you see in Figure 8-2. Each
peak represents a different base. The sequence indicated by the peaks is the
complement of the template strand (see Chapter 6 for more on the complementary nature of DNA). When you know the complement of the template,
you know the template sequence itself. You can then mine this information
for the location of genes (see Chapter 10) and compare it to the sequences of
other organisms, such as those listed in Table 8-1.

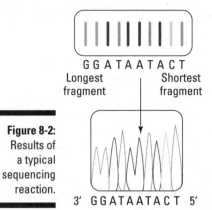

GGATAATACT

Longest fragment   Shortest fragment

Figure 8-2: Results of a typical sequencing reaction.

3′ GGATAATACT 5′

Open access and the Human Genome Project

Prior to the Human Genome Project (HGP), sequencing was a very difficult and time-consuming enterprise. Getting a 1,000-base long sequence required about three days of work and used radioactive chemicals instead of dyes. Sequences were read by hand and had to be run over and over again to fill in gaps and correct mistakes. Every single sequence had to be entered into the computer by hand — imagine typing thousands of As, Gs, Ts, and Cs! It would have taken centuries to sequence the human genome using the old methods. The sheer magnitude of the HGP required faster and easier techniques.

Numerous companies, government labs, and universities searched for solutions to make sequencing faster, better, and cheaper. When the HGP began, one automated sequencer machine produced 1,500 sequences (of 1,000 base pairs each) in about 24 hours. Many laboratories worked together using automated sequencers running 24 hours a day to power through the entire human genome. Still, it took about 15 years to complete the HGP!

New technologies are leaving the once-grand HGP in the dust, as sequencing entire genomes is faster and cheaper than ever. For example, a microbe genome that required 3 months of work in 1995 was entirely sequenced in just 4 hours in 2006. Using high-throughput sequencing, it takes a staff of four people about one month to decode an entire human's genome at a cost of $50,000 (as compared to the original HGP, which came in at roughly $500 million). These new technologies are paving the way for personalized medicine, rapid detection of disease, gene therapy, and much more.

Chapter 9

RNA: DNA's Close Cousin

DNA is the stuff of life. Practically every organism on earth relies on DNA to store genetic information and transmit it from one generation to the next. The road from *genotype* (building plans) to *phenotype* (physical traits) begins with *transcription* — making a special kind of copy of DNA. DNA is so precious and vital to *eukaryotes* (organisms made up of cells with nuclei) that it's kept packaged in the cell nucleus, like a rare document that's copied but never removed from storage. Because it can't leave the safety of the nucleus, DNA directs all the cell's activity by delegating responsibility to another chemical, RNA. RNA carries messages out of the cell nucleus into the cytoplasm (visit Chapter 2 for more about navigating the cell) to direct the production of proteins during *translation,* a process you find out more about in Chapter 10.

You Already Know a Lot about RNA

If you read Chapter 6, in which I cover DNA at length, you already know a lot about *ribonucleic acid,* or RNA. From a chemical standpoint, RNA is very simple. It's composed of

✔ Ribose sugar (instead of deoxyribose, which is found in DNA)

✔ Four nucleotide bases (three you know from DNA — adenine, guanine, and cytosine — plus an unfamiliar one called *uracil*)

✔ Phosphate (the same phosphate found in DNA)

RNA has three major characteristics that make it different from DNA:

✔ RNA is very unstable and decomposes rapidly.

✔ RNA contains uracil in place of thymine.

✔ RNA is almost always single-stranded.

Using a slightly different sugar

Both RNA and DNA use a *ribose* sugar as a main element of their chemical structures. The ribose sugar used in DNA is deoxyribose (find out more about this sugar in Chapter 6). RNA, on the other hand, uses unmodified ribose. Take a careful look at Figure 9-1. You can see that three spots on ribose are marked with numbers. (On ribose sugars, numbers are followed by an apostrophe ['] to indicate the designation "prime;" see Chapter 6 for more information.) Ribose and deoxyribose both have an oxygen (O) atom and a hydrogen (H) atom (an OH group) at their 3' sites.

OH groups are also called *reactive groups* because oxygen atoms are very aggressive from a chemical standpoint (so aggressive that some chemists say they "attack" incoming atoms). The 3' OH tail is required for phosphodiester bonds to form between nucleotides in both ribose and deoxyribose atoms, thanks to their aggressive oxygen atoms. (For the scoop on how phosphodiester bonds form during replication, see Chapter 7.)

Figure 9-1:
The ribose
sugar is part
of RNA.

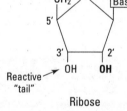

The difference between the two molecules is an oxygen atom at the 2' spot: absent (with deoxyribose) or present (with ribose). This one oxygen atom has a huge hand in the differing purposes and roles of DNA and RNA:

✔ **DNA:** DNA must be protected from decomposition. The absence of one oxygen atom is part of the key to extending DNA's longevity. When the 2' oxygen is missing, as in deoxyribose, the sugar molecule is less likely to get involved in chemical reactions (because oxygen is chemically aggressive); by being aloof, DNA avoids being broken down.

✔ **RNA:** RNA easily decomposes because its reactive 2' OH tail introduces RNA into chemical interactions that break up the molecule. Unlike DNA, RNA is a short-term tool the cell uses to send messages and manufacture proteins as part of gene expression (which I cover in Chapter 10). Messenger RNAs (mRNAs) carry out the actions of genes. Put simply, to turn a gene "on," mRNAs have to be made, and to turn a gene "off," the mRNAs that turned it "on" have to be removed. So the 2' OH tail is a built-in mechanism that allows RNA to be decomposed, or *removed,* rapidly and easily when the message is no longer needed and the gene needs to be turned "off" (see Chapter 11 for more on turning genes off and on).

Meeting a new base: Uracil

RNA is composed of four nucleotide bases. Three of the four bases may be quite familiar to you because they're also part of DNA: adenine (A), guanine (G), and cytosine (C). The fourth base, uracil (U), is found only in RNA. (In DNA, the fourth base is thymine. See Chapter 6 for details.) RNA's bases are pictured in Figure 9-2.

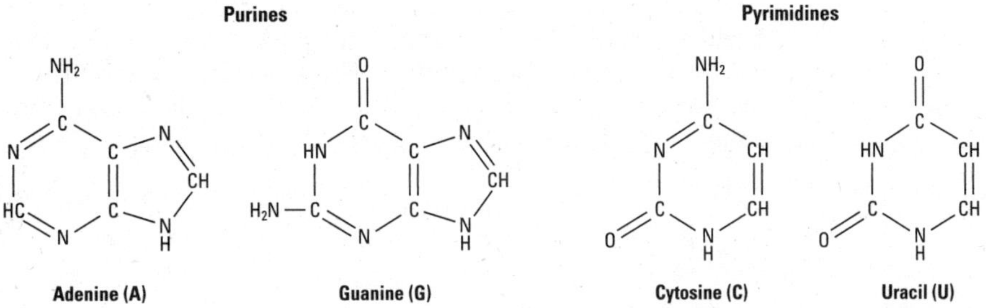

Purines   **Pyrimidines**

Figure 9-2: The four bases found in RNA.

Adenine (A)   Guanine (G)   Cytosine (C)   Uracil (U)

Uracil may be new to you, but it's actually the precursor of DNA's thymine. When your body produces nucleotides, uracil is hooked up with a ribose and three phosphates to form a ribonucleoside triphosphate (rNTP). (Check out Figure 9-5 later in the chapter to see an rNTP.) If DNA is being replicated, or copied (see Chapter 7 for the details on DNA's copying process), deoxyribonucleotide triphosphates (dNTPs) of thymine — not uracil — are needed, meaning that a few things have to happen:

✔ The 2' oxygen must be removed from ribose to make deoxyribose.

✔ A chemical group must be added to uracil's ring structure (all the bases are rings; see Chapter 6 for details on how these rings stack up). Folic acid, otherwise known as vitamin B9, helps add a carbon and three hydrogen atoms (CH_3, referred to as a *methyl group*) to uracil to convert it to thymine.

Uracil carries genetic information in the same way thymine does, as part of sequences of bases. (In fact, the genetic code that's translated into protein is written using uracil; see Chapter 10 for more on the genetic code.)

The complementary base pairing rules that apply to DNA (see Chapter 6) also apply to RNA: purines with pyrimidines (that is, G with C and A with U). So why are there two versions of essentially the same base (uracil and thymine)?

- ✔ Thymine protects the DNA molecule better than uracil can because that little methyl group (CH_3) helps make DNA less obvious to chemicals called *nucleases* that chew up both DNA and RNA. Nucleases are *enzymes* (chemicals that cause reactions to occur) that act on nucleic acids (see Chapter 6 for why DNA and RNA are called nucleic acids). Your body uses nucleases to attack unwanted RNA and DNA molecules (such as viruses and bacteria), but if methyl groups are present, nucleases can't bond as easily with the nucleic acid to break its chains. (The methyl group also makes DNA hydrophobic; see Chapter 6 for why DNA is afraid of water.)

- ✔ Uracil is a very friendly base; it easily bonds with the other three bases to form pairs. Uracil's amorous nature is great for RNA, which needs to form all sorts of interesting turns, twists, and knots to do its job (see the next section, "Stranded!"). DNA's message is too important to trust to such an easygoing base as uracil; strict base pairing rules must be followed to protect DNA's message from mutation (see Chapter 13 for more on how base pair rules protect DNA's message from getting garbled). Thymine, as uracil's less friendly near-twin, only bonds with adenine, making it perfectly suited to protect DNA's message.

Stranded!

RNA is almost always single-stranded, and DNA is always double-stranded. The double-stranded nature of DNA helps protect its message and provides a simple way for the molecule to be copied during replication. Like DNA, RNA loves to hook up with complementary bases. But RNA is a bit narcissistic; it likes to form bonds with itself (see Figure 9-3), creating what's called a *secondary structure*. The primary structure of RNA is the single-stranded molecule; when the molecule bonds with itself and gets all twisted and folded up, the result is the secondary structure.

Three major types of RNA carry out the business of expressing DNA's message (Chapter 23 covers other RNAs). Although all three RNAs function as a team during translation (which I cover in Chapter 10), the individual types carry out very specific functions.

✔ **mRNA:** Carries out the actions of genes

✔ **tRNA:** Carries amino acids around during translation (see Chapter 10 for more on translation)

✔ **rRNA:** Puts amino acids together in chains (see Chapter 10 for more on rRNA's role during translation)

Figure 9-3:
Single-
stranded
RNAs form
interesting
shapes in
order to
carry out
various
functions.

Primary Structure

5′ | AUGCGGCUACGUAACGAGCUUAGCGCGUAUACCGAAAGGGUAGAAC | 3′

Complementary regions bond
to form secondary structure

Transcription: Copying DNA's Message into RNA's Language

A *transcript* is a record of something, not an exact copy. In genetics, *transcription* is the process of recording part of the DNA message in a related, but different, language — the language of RNA. (To review differences between DNA and RNA, jump back to "You Already Know a Lot about RNA," earlier in this chapter.) Transcription is necessary because DNA is too valuable to be moved or tampered with. The DNA molecule is *the* plan, and any error that's introduced into the plan (as a mutation, which I address in Chapter 13) causes lots of problems. If part or all of the DNA molecule were lost, the cell would die (flip to Chapter 14 for more on cell death). Transcription keeps DNA safe by letting a temporary RNA copy take the risk of leaving the cell nucleus and going out into the cytoplasm.

Messenger RNAs (mRNAs) are the specific type of RNA responsible for carrying DNA's message from the cell nucleus into the cytoplasm (check out Chapter 2 for a review of cell parts).

With *transcription,* the DNA inside the nucleus goes through a process similar to *replication* (see Chapter 7) to get the message out as RNA. When DNA is replicated, the result is another DNA molecule that's exactly like the original in every way. But in transcription, many mRNAs are created because, instead of transcribing the entire DNA molecule, only messages of genes are transcribed into mRNA. Transcription has several steps:

1. Enzymes identify the right part of the DNA molecule to transcribe (see the upcoming section "Getting ready to transcribe").

2. The DNA molecule is opened up to make the message accessible (see "Initiation").

3. Enzymes build the mRNA strand (see "Elongation").

4. The DNA molecule snaps shut to release the newly synthesized mRNA (see "Termination").

Getting ready to transcribe

In preparing to transcribe DNA into mRNA, three things need to be completed:

- ✔ Locate the proper gene sequence within the billions of bases that make up DNA.
- ✔ Determine which of the two strands of DNA to transcribe.
- ✔ Gather up the nucleotides of RNA and the enzymes needed to carry out transcription.

Locating the gene

Your chromosomes are made up of roughly 3 billion base pairs of DNA and contain roughly 22,000 genes (see Chapter 8). But only about 1 percent of your DNA gets transcribed into mRNA. Genes, the sequences that do get transcribed, vary in size. The average gene is only about 3,000 base pairs long, but the human genome also has some gigantic genes — for example, the gene that's implicated in a particular form of muscular dystrophy (Duchenne) is a whopping 2.5 million base pairs.

Before a gene of any size can be transcribed, it must be located. The cue that says "start transcription here" is written right into the DNA in regions called *promoters.* (The promoter also controls how often the process takes place; see the "Initiation" section later in the chapter.) The sequence that indicates where to stop transcribing is called a *terminator.* The gene, the promoter, and the terminator together are called the *transcription unit* (see Figure 9-4).

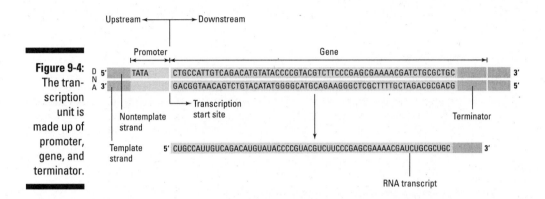

Figure 9-4: The transcription unit is made up of promoter, gene, and terminator.

The promoter sequences tell the enzymes of transcription where to start work and are located within 30 or so base pairs of the genes they control. Each gene has its own promoter. In eukaryotes, the beginning sequence of the promoter is always the same, and it's called the *TATA box* because the sequence of the bases is TATAAA. The presence of TATA tells the transcription-starting enzyme that the gene to transcribe is about 30 base pairs away. Sequences like TATA that are the same in many (if not all) organisms are called *consensus sequences,* indicating that the sequences agree or mean the same thing everywhere they appear.

Locating the right strand

By now you've (hopefully) picked up on the fact that DNA is double-stranded. Those double strands aren't identical, though; they're complementary, meaning that the sequence of bases matches up, but it doesn't spell the same words of the genetic code (see Chapter 10 for genetic code info). The genetic code of DNA works like this: Bases of genes are read in three base sets, like words. For example, three adenines in a row (AAA) are transcribed into mRNA as three uracils (UUU). During translation, UUU tells the ribosome to use an amino acid called phenylalanine as part of the protein it's making. If the complementary DNA, TTT, were transcribed, you'd wind up with an mRNA saying AAA, which specifies lysine. A protein containing lysine will function differently than one containing phenylalanine.

Because complements don't spell the same genetic words, you can get two different messages depending on which strand of DNA is transcribed into mRNA. Therefore, genes can only be read from *one* of the two strands of the double-stranded DNA molecule — but which one? The TATA box (the promoter; see the preceding "Locating the gene" section) not only indicates where a gene is but also tells which strand holds the gene's information. TATA boxes indicate that a gene is about 30 bases away going in the 3' direction (sometimes referred to as *downstream*). Genes along the DNA molecule run in both directions, but any given gene is transcribed only in the 3' direction. Because only one strand is transcribed, the two strands are designated in one of two ways:

  ✔ **Template:** This strand provides the pattern for transcription.

  ✔ **Nontemplate:** This strand is the original message that's actually being transcribed.

TATA is on the nontemplate strand and indicates that the other (complementary) strand is to be used as the template for transcription. Look back at Figure 9-4 and compare the template to the RNA transcript — they're complementary. Now compare the mRNA transcript to the nontemplate strand. The only difference between the two is that uracil appears in place of thymine. The RNA is the transcript of the nontemplate strand.

Gathering building blocks and enzymes

In addition to template DNA (see the preceding section), the following ingredients are needed for successful transcription:

  ✔ **Ribonucleotides,** the building blocks of RNA

  ✔ **Enzymes and other proteins,** to assemble the growing RNA strand in the process of *RNA synthesis*

The building blocks of RNA are nearly identical to those used in DNA synthesis, which I explain in Chapter 7. The differences, of course, are that for RNA, ribose is used in place of deoxyribose, and uracil replaces thymine. Otherwise, the rNTPs (ribonucleoside triphosphates; see Figure 9-5) look very much like the dNTPs you're hopefully already familiar with.

In a process similar to replication, transcription requires the services of various enzymes to:

  ✔ Find the promoter (see the earlier "Locating the gene" section)

  ✔ Open up the DNA molecule (see the later "Initiation" section)

  ✔ Assemble the growing strand of RNA (see the later "Elongation" section)

Unlike replication, though, transcription has fewer enzymes to keep track of. The main player is RNA polymerase. Like DNA polymerase (which you can meet in Chapter 7), *RNA polymerase* recognizes each base on the template and adds the appropriate complementary base to the growing RNA strand, substituting uracil where DNA polymerase would supply thymine. RNA polymerase hooks up with a large group of enzymes — called a *holoenzyme* — to carry out this process. The individual enzymes making up the holoenzyme vary between prokaryotes and eukaryotes, but their functions remain the same: to recognize and latch onto the promoter and to call RNA polymerase over to join the party.

Eukaryotes have three kinds of RNA polymerase, which vary only in which genes they transcribe.

✔ RNA polymerase I takes care of long rRNA molecules.

✔ RNA polymerase II carries out the synthesis of most mRNA and some tiny, specialized types of RNA molecules that are used in RNA editing after transcription is over (see "Post-transcription Processing" later in this chapter).

✔ RNA polymerase III transcribes tRNA genes and other small RNAs used in RNA editing.

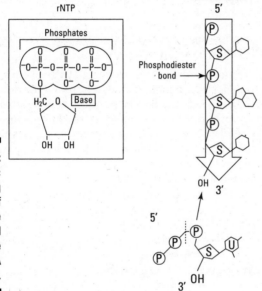

Figure 9-5:
The basic building blocks of RNA and the chemical structure of an RNA strand.

Initiation

Initiation includes finding the gene and opening up the DNA molecule so that the enzymes can get to work. The process of initiation is pretty simple:

1. **The holoenzyme (group of enzymes that hook up with RNA polymerase) finds the promoter.**

   The promoter of each gene controls how often transcription makes an mRNA transcript to carry out the gene's action. RNA polymerase can't bind to a gene that isn't scheduled for transcription. In eukaryotes, _enhancers,_ which are sequences sometimes distantly located from the transcription unit, also control how often a particular gene is transcribed. To find out more about how genes are turned on, flip to Chapter 11.

2. RNA polymerase opens up the double-stranded DNA molecule to expose a very short section of the template strand.

When the promoter "boots up" to initiate transcription, the holoenzyme complex binds to the promoter site and signals RNA polymerase. RNA polymerase binds to the template at the start site for transcription. RNA polymerase can't "see" past the sugar-phosphate backbone of DNA, so transcription can't occur if the molecule isn't first opened up to expose single strands. RNA polymerase melts the hydrogen bonds between the double-stranded DNA molecule and opens up a short stretch of the helix to expose the template. The opening created by RNA polymerase when it wedges its way between the two strands of the helix is called the *transcription bubble* (see Figure 9-6).

3. RNA polymerase strings together rNTPs to form mRNA (or one of the other types of RNA, such as tRNA or rRNA).

RNA polymerase doesn't need a primer to begin synthesis of a new mRNA molecule (unlike DNA replication; see Chapter 7 for details). RNA polymerase simply reads the first base of the transcription unit and lays down the appropriate complementary rNTP. This first rNTP doesn't lose its three phosphate molecules because no phosphodiester bond is formed at the 5' side. Those two extra phosphates remain until the mRNA is edited later in the transcription process (see "Post-transcription Processing" later in this chapter).

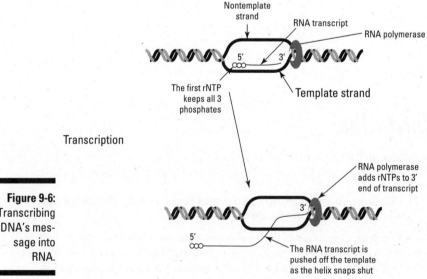

Nontemplate strand

RNA transcript

RNA polymerase

The first rNTP keeps all 3 phosphates

Template strand

Transcription

RNA polymerase adds rNTPs to 3' end of transcript

Figure 9-6: Transcribing DNA's message into RNA.

The RNA transcript is pushed off the template as the helix snaps shut

Elongation

After RNA polymerase puts down the first rNTP, it continues opening the DNA helix and synthesizing mRNA by adding rNTPs until the entire transcriptional unit is transcribed. The transcription bubble (the opening between DNA strands) itself is very small; only about 20 bases of DNA are exposed at a time. So as RNA polymerase moves down the transcription unit, only the part of the template that's actively being transcribed is exposed. The helix snaps shut as RNA polymerase steams ahead to push the newly synthesized mRNA molecule off the template (refer to Figure 9-6). An enzyme like gyrase (see Chapter 7) probably works to keep the DNA molecule from getting knotted up during the opening, transcribing, and closing process (but scientists aren't certain at this point).

The transcriptional units of genes contain sequences that aren't translated into protein. However, these sequences may control how genes are expressed (see Chapter 11 to find out more). As you may expect, geneticists have come up with terms for the parts that are translated and those that aren't:

- ✔ **Introns:** Noncoding sequences that get their name from their *in*tervening presence. Genes often have many introns that fall between the parts of the gene that code for phenotype.
- ✔ **Exons:** Coding sequences that get their name from their *ex*pressed nature.

The entire gene — introns and exons — is transcribed (refer to Figure 9-6). After transcription has terminated, part of the editing process is the removal of introns. I cover the process of snipping out introns and splicing together exons in the section "Editing the message," later in this chapter.

Prokaryotes don't have introns because prokaryotic genes are all coding, or exon. Only eukaryotes have genes interrupted by intron sequences. Almost all eukaryotic genes have at least one intron; the maximum number of introns in any one gene is 200. Scientists continue to explore the function of introns, which in part control how different mRNAs are edited.

Termination

When RNA polymerase encounters the terminator (as a sequence in the DNA, not the scary, gun-toting movie character), it transcribes the terminator sequence and then stops transcription. What happens next varies depending on the organism.

    ✔ In prokaryotic cells, some terminator sequences have a series of bases that are complementary and cause the mRNA to fold back on itself. The folding stops RNA polymerase from moving forward and pulls the mRNA off the template.

    ✔ In eukaryotic cells, a special protein called a *termination factor* aids RNA in finding the right stopping place.

In any event, after RNA polymerase stops adding rNTPs, the mRNA gets detached from the template. The holoenzyme and RNA polymerase let go of the template, and the double-stranded DNA molecule snaps back into its natural helix shape.

Post-transcription Processing

Before mRNA can venture out of the cell nucleus and into the cytoplasm for translation, it needs a few modifications. And I just happen to cover them in the following sections.

Adding cap and tail

The "naked" mRNA that's produced by transcription needs to get dressed before translation:

    ✔ A 5' cap is added.

    ✔ A long tail of adenine bases is tacked on.

RNA polymerase starts the process of transcription by using an unmodified rNTP (see the section "Initiation" earlier in this chapter). But a 5' cap needs to be added to the mRNA to allow the ribosome to recognize it during translation (see Chapter 10 for more on translation). The first part of adding the cap is the removal of one of the three phosphates from the leading end of the mRNA strand. A guanine, in the form of a ribonucleotide, is then attached to the lead base of the mRNA. (Figure 9-7 illustrates the process of cap and tail attachment to the mRNA.) Several groups composed of a carbon atom with three hydrogen atoms (CH_3, called a *methyl group*) attach at various sites — on the guanine and on the first and second nucleotides of the mRNA. Like the methyl groups that protect the thymine-bearing DNA molecule, the methyl groups at the 5' end of the mRNA protect it from decomposition and allow the ribosome to recognize the mRNA as ready for translation.

In eukaryotes, a long string of adenines is added to the 3' end of the mRNA to further protect the mRNA from natural nuclease activity long enough to get translated (see Figure 9-7). This string is called the *poly-A tail*. RNA molecules are easily degraded and destroyed because of their temporary natures. Like memos, RNA molecules are linked to a specific task, and when the task is over, the memo is discarded. But the message has to last long enough to be read, sometimes more than once, before it hits the shredder (in this case, nucleases do the shredding instead of guilty business executives). The length of the poly-A tail determines how long the message lasts and how many times it can be translated by the ribosomes before nucleases eat the tail and destroy the message.

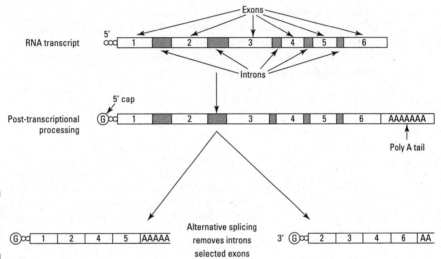

Figure 9-7:
Capping
things off.

Editing the message

The final step in preparing mRNA for translation is twofold: removing the noncoding intron sequences and stringing the exons together without inter-ruptions between them. Several specialized types of RNA work to find the start and end points of introns, pull the exons together, and snip out the extra RNA (that is, the intron).

While it's still in the nucleus, a complex of proteins and small RNA molecules called a *spliceosome* inspects the newly manufactured mRNA. The spliceo-some is like a roaming workshop that recognizes introns and works to remove them from between exons. The spliceosome recognizes consensus sequences that mark the beginnings and endings of introns (look back at the "Locating the gene" section to review consensus sequences). The spliceosome grabs each end of the intron and pulls the ends toward each other to form a loop. This movement has the effect of bringing the beginning of one exon close to the end of the preceding one. The spliceosome then snips out the intron and hooks the exons together in a process called *splicing*. Splicing creates a phosphodiester bond between the two exon sequences, which seals them together as one strand of mRNA.

Introns can be spliced out leaving all the exons in their original order, or introns *and* exons can be spliced out to create a new sequence of exons (refer to Figure 9-6 for a couple of examples). The splicing of introns and exons is called *alternative splicing* and results in the possibility for one gene to be expressed in different ways. Thanks to alternative splicing, the 22,000 or so genes in humans are able to produce around 90,000 different proteins. New evidence suggests that practically all multi-exon genes (which make up roughly 86 percent of the human genome) can be sliced and diced in multiple ways, thanks to alternative splicing.

One of the secrets to the genetic flexibility of alternative splicing is sequences called *Alu elements*. Alu elements are fairly short sequences that show up all over the human genome (see Chapter 8 for how scientists are exploring the human genome) — your DNA may have as many as 1 million copies of Alu. Alu can be spliced into or out of genes (sometimes more than once) to create alternative forms of mRNA from the same original gene sequence. This sequence turns out to act as an exon and is considered a reason to scrap the term "junk DNA" altogether. The enormous versatility of RNA editing has lead some scientists to think of RNA as "the" genetic material instead of DNA (see Chapter 23).

After the introns are spliced and all the exons are strung together, the mRNA molecule is complete and ready for action. It migrates out of the cell nucleus, encounters an army of ribosomes, and goes through the process of translation — the final step in converting the genetic message from DNA to protein.

Chapter 10

Translating the Genetic Code

. . . .

In This Chapter

▶ Exploring the features of the genetic code

▶ Translating genetic information into phenotype

▶ Molding polypeptides into functional proteins

. . . .

From building instructions to implementation, the message that DNA carries follows a predictable path. First, DNA provides the template for transcription of the message into RNA. Then, RNA (in the form of messenger RNA) moves out of the cell nucleus and into the cytoplasm to provide the building plans for *proteins*. Every living thing is made of proteins, which are long chains of amino acids called *polypeptides* that are folded into complex shapes and hooked together in intricate ways.

All the physical characteristics (that is, the *phenotypes*) of your body are made up of thousands of different proteins. Of course, your body is also composed of other things, too, like water, minerals, and fats. But proteins supply the framework to organize all those other building blocks, and proteins carry out all your necessary bodily functions, like digestion, respiration, and elimination.

In this chapter, I explain how RNA provides the blueprint for manufacturing proteins, the final step in the transformation from *genotype* (genetic information) to phenotype. Before you dive into the translation process, you need to know a few things about the genetic code — the information that mRNA carries — and how the code is read. If you skipped over Chapter 8, you may want to go back and review its material on RNA before moving on.

Discovering the Good in a Degenerate

When Watson and Crick (along with Rosalind Franklin; see Chapter 6 for the full scoop) discovered that DNA is made up of two strands composed of four bases, the big question they faced was: How can only four bases contain enough information to encode complex phenotypes?

Complex phenotypes (such as your bone structure, eye color, and ability to digest spicy food) are the result of combinations of proteins. The genetic code (that is, DNA transcribed as RNA; see Chapter 9) provides the instructions to make these proteins (via translation; see "Meeting the Translating Team" later in this chapter). Proteins are made up of amino acids strung together in various combinations to create chains called polypeptides (which is a fancy way of saying "protein"). Polypeptide chains can vary from 50 to 1,000 amino acids in length. Because there are 20 different amino acids, and because chains are often more than 100 amino acids in length, the variety of combinations is enormous. For example, a polypeptide that's only 5 amino acids long has 3,200,000 combinations!

After experiments showed that DNA was truly the genetic material (see Chapter 6), skeptics continued to point to the simplicity of the four bases in RNA and argued that a code of four bases wouldn't work to encode complex peptides. Reading the genetic code one base at a time — U, C, A, and G — would mean that there simply weren't enough bases to make 20 amino acids. So it was obvious to scientists that the code must be made up of multiple bases read together. A two-base code didn't work because it only produced 16 combinations — too few to account for 20 amino acids. A three-base code (referred to as a *triplet code*) looked like overkill, because a *codon,* which is a combination of three nucleotides in a row, that chooses from four bases at each position produces 64 possible combinations. Skeptics argued that a triplet code contains too much redundancy — after all, there are only 20 amino acids.

As it turns out, the genetic code is *degenerate,* which is a fancy way of saying "too much information." Normally, degenerate means something to the effect of "bad and getting worse" (it's usually used to describe some people — I won't name names). In the genetic sense, the degeneracy of the triplet code means that the code is highly flexible and tolerates some mistakes — which is a good thing.

Several features of the genetic code are important to keep in mind. The code is

- ✔ **Triplet,** meaning that bases are read three at a time in codons.

- ✔ **Degenerate,** meaning that 18 of the 20 amino acids are specified by two or more codons (see the next section, "Considering the combinations").

- ✔ **Orderly,** meaning that each codon is read in only one way and in only one direction, just as English is read left to right (see "Framed! Reading the code" later in this chapter).

- ✔ **Nearly universal,** meaning that just about every organism on earth interprets the language of the code in exactly the same way (see "Not quite universal" for exceptions).

Considering the combinations

Only 61 of the 64 codons are used to specify the 20 amino acids found in proteins. The three codons that don't code for any amino acid simply spell "stop," telling the ribosome to cease the translation process (see "Termination" later in this chapter). In contrast, the one codon that tells the ribosome that an mRNA is ripe for translating — the "start" codon — codes for an amino acid, *methionine*. (The "start" amino acid comes in a special form; see "Initiation" later in this chapter.) In Figure 10-1, you can see the entire code with all the alternative spellings for the 20 amino acids. (See "Meeting the Translating Team" later in this chapter for more details about amino acids.)

First Letter ↓	Second Letter				Third Letter ↓
	U	C	A	G	
U	phenylalanine	serine	tyrosine	cysteine	U
	phenylalanine	serine	tyrosine	cysteine	C
	leucine	serine	STOP	STOP	A
	leucine	serine	STOP	tryptophan	G
C	leucine	proline	histidine	arginine	U
	leucine	proline	histidine	arginine	C
	leucine	proline	glutamine	arginine	A
	leucine	proline	glutamine	arginine	G
A	isoleucine	threonine	asparagine	serine	U
	isoleucine	threonine	asparagine	serine	C
	isoleucine	threonine	lysine	arginine	A
	methionine & START	threonine	lysine	arginine	G
G	valine	alanine	aspartate	glycine	U
	valine	alanine	aspartate	glycine	C
	valine	alanine	glutamate	glycine	A
	valine	alanine	glutamate	glycine	G

Figure 10-1: The 64 codons of the genetic code, as written by mRNA.

For many of the amino acids, the alternative spellings differ only by one base — the third base of the codon. For example, four of the six spellings for leucine start with the bases CU. This flexibility at the third position of the codon is called a *wobble*. The third base of the mRNA can vary, or wobble, without changing the meaning of the codon (and thus the amino acid it codes for). The wobble is possible because of the way *tRNAs (transfer RNAs)* and mRNAs pair up during the process of translation. The first two bases of the code on the mRNA and the partner tRNA (which is carrying the amino acid specified by the codon) must be exact matches. However, the third base of the tRNA can break the base-pairing rules, allowing bonds with mRNA bases other than the usual complements. This rule violation, or wobble, allows different spellings to code for the same amino acid. However, some codons, like one of the three stop codons (spelled UGA), have only one meaning; wobbles in this stop codon change the meaning from stop to either cysteine (spelled UGU or UGC) or tryptophan (UGG).

Framed! Reading the code

Besides its combination possibilities, another important feature of the genetic code is the way in which the codons are read. Each codon is separate, with no overlapping. And the code doesn't have any punctuation — it's read straight through without pauses.

The codons of the genetic code run sequentially, as you can see in Figure 10-2. Each codon is read only once using a *reading frame,* a series of sequential, non-overlapping codons. The start codon defines the position of the reading frame. In the mRNA in Figure 10-2, the sequence AUG, which spells methionine, is a start codon. After the start codon, the bases are read three at a time without a break until the stop codon is reached. (Mutations often disrupt the reading frame by inserting or removing one base; see Chapter 13 for more details.)

Figure 10-2:
The genetic code is non-overlapping and uses a reading frame.

Nucleotide sequence

A U G C G A G U C U U G C A G . . .

Nonoverlapping code

A U G C G A G U C U U G C A G . . .
　1　　2　　3　　4　　5

Not quite universal

The meaning of the genetic code is nearly universal. That means nearly every organism on earth uses the same spellings in the triplet code. Mitochondrial DNA spells a few words differently from nuclear DNA, which may explain (or at least relate back to) mitochondria's unusual origins (see Chapter 6). Plants, bacteria, and a few microorganisms also use unusual spellings for one or more amino acids. Otherwise, the way the code is read — influenced by its degenerate nature, with wobbles, without punctuation, and using a specific reading frame — is the same. As scientists tackle DNA sequencing for various creatures (see Chapter 8), more unusual spellings are likely to pop up.

Meeting the Translating Team

Translation is the process of converting information from one language into another. In this case, the genetic language of nucleic acid is translated into the language of protein. Translation takes place in the cytoplasm of cells. After messenger RNAs (mRNAs) are created through transcription and move into the cytoplasm, the protein production process begins (see Chapter 9 for the lowdown on mRNA). The players involved in protein production include

- **Ribosome:** The big protein-making factory that reads mRNA's message and carries out the message's instructions. Ribosomes are made up of *ribosomal RNA* (rRNA) and are capable of constructing any sort of protein.

- **The genetic code:** The message carried by mRNA (see "Discovering the Good in a Degenerate" earlier in this chapter for more on the genetic code).

- **Amino acids:** Complex chemical compounds containing nitrogen and carbon; 20 amino acids strung together in thousands of unique combinations are used to construct proteins.

- **Transfer RNA (tRNA):** Runs a courier service to provide amino-acid building blocks to the working ribosome; each tRNA summoned by the ribosome grabs the amino acid that the codon specifies.

Taking the Translation Trip

Translation proceeds in a series of predictable steps:

1. A ribosome recognizes an mRNA and latches onto its 5' cap (see Chapter 8 for an explanation of how and why mRNAs get caps). The ribosome slurps up the mRNA and carefully scrutinizes it, looking for codons that form the words of the genetic code, beginning with the start codon.

2. tRNAs supply the amino acids dictated by each codon when the ribosome reads the instructions. The ribosome assembles the polypeptide chain with the help of various enzymes and other proteins.

3. The ribosome continues to assemble the polypeptide chain until it reaches the stop codon. The completed polypeptide chain is released.

After it's released from the ribosome, the polypeptide chain is modified and folded to become a mature protein.

Initiation

Preparation for translation consists of two major events:

- ✔ The tRNA molecules must be hooked up with the right amino acids in a process called *charging.*
- ✔ The ribosome, which comes in two pieces, must assemble itself at the start codon of the mRNA.

Charge! tRNA hooks up with a nice amino acid

Transfer RNA (tRNA) molecules are small, specialized RNAs produced by transcription. However, unlike mRNAs, tRNAs are never translated into protein; tRNA's whole function is ferrying amino acids to the ribosomes for assembly into polypeptides. tRNAs are uniquely shaped to carry out their job. In Figure 10-3, you see two depictions of tRNA. The illustration on the left shows you tRNA's true form. The illustration on the right is a simplified version that makes tRNA's parts easier to identify. The cloverleaf shape is one of the keys to the way tRNA works. tRNA gets its unusual configuration because many of the bases in its sequence are complements; the strand folds, and the complementary bases form bonds, resulting in the loops and arms of a typical tRNA.

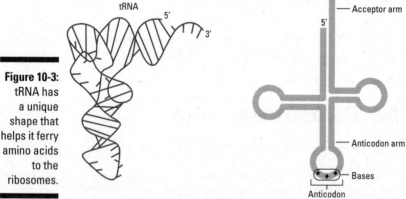

Figure 10-3: tRNA has a unique shape that helps it ferry amino acids to the ribosomes.

The two key elements of tRNA are

- ✔ **Anticodon:** A three-base sequence on one loop of each tRNA; the anticodon is complementary to one of the codons spelled by mRNA.
- ✔ **Acceptor arm:** The single-stranded tail of the tRNA; where the amino acid corresponding to the codon is attached to the tRNA.

The codon of mRNA specifies the amino acid used during translation. The anti-codon of the tRNA is complementary to the codon of mRNA and specifies which amino acid each tRNA is built to carry.

Like a battery, tRNAs must be charged in order to work. tRNAs get charged with the help of a special group of enzymes called *aminoacyl-tRNA synthetases.* Twenty synthetases exist, one for each amino acid specified by the codons of mRNA. Take a look at the illustration on the right in Figure 10-3, the schematic of tRNA. The aminoacyl-tRNA synthetases recognize sequences of bases in the anticodon of the tRNA that announce which amino acid that particular tRNA is built to carry. When the aminoacyl-tRNA synthetase encounters the tRNA molecule that matches its amino acid, the synthetase binds the amino acid to the tRNA at the acceptor arm — this is the charging part. Figure 10-4 shows the connection of amino acid and tRNA. The synthetases proofread to make sure that each amino acid is on the appropriate tRNA. This proofreading ensures that errors in tRNA charging are very rare and prevents errors in translation later on. With the amino acid attached to it, the tRNA is charged and ready to make the trip to the ribosome.

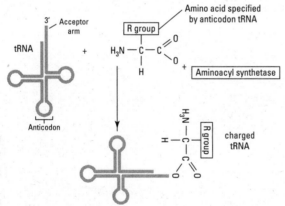

Figure 10-4:
tRNA
charging.

Putting the ribosome together

Ribosomes come in two parts called *subunits* (see Figure 10-5), and ribosomal subunits come in two sizes: large and small. The two subunits float around (sometimes together and sometimes as separate pieces) in the cytoplasm until translation begins. Unlike tRNAs, which match specific codons, ribosomes are completely flexible and can work with any mRNA they encounter. Because of their versatility, ribosomes are sometimes called "the workbench of the cell."

When fully assembled, each ribosome has two sites and one slot:

✔ **A-site (acceptor site):** Where tRNA molecules insert their anticodon arms to match up with the codon of the mRNA molecule.

✔ **P-site (peptidyl site):** Where amino acids get hooked together using peptide bonds.

✔ **Exit slot:** Where tRNAs are released from the ribosome after their amino acids become part of the growing polypeptide chain.

Before translation can begin, the smaller of the two ribosome subunits attaches to the 5' cap of the mRNA with the help of proteins called *initiation factors*. The small subunit then scoots along the mRNA until it hits the start codon (AUG). The P-site on the small ribosome subunit lines up with the start codon, and the small subunit is joined by the tRNA carrying methionine (UAC), the amino acid that matches the start codon. The "start" tRNA totes a special version of methionine called *fMet* (short for *N-formylmethionine*). Only the tRNA for fMet can attach to the ribosome at the P-site without first going through the A-site. The tRNA uses its anticodon, which is complementary to the codon of the mRNA, to hook up to the mRNA. The large ribosome subunit joins with the small subunit to begin the process of hooking together all the amino acids specified by the mRNA (refer to Figure 10-5).

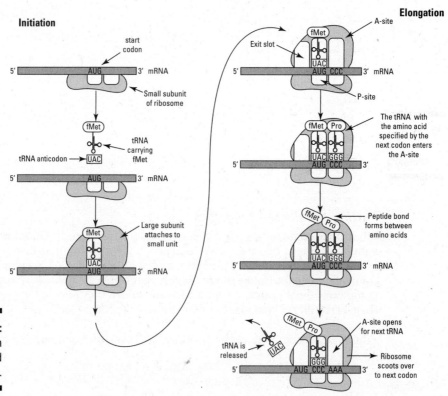

Figure 10-5:
Initiation and elongation.

Elongation

When the initiation process is complete, translation proceeds in several steps called *elongation,* which you can follow in Figure 10-5.

1. The ribosome calls for the tRNA carrying the amino acid specified by the codon residing in the A-site. The appropriate charged tRNA inserts its anticodon arm into the A-site.

2. Enzymes bond the two amino acids attached to the acceptor arms of the tRNAs in the P- and A-sites.

3. As soon as the two amino acids are linked together, the ribosome scoots over to the next codon of the mRNA. The tRNA that was formerly in the P-site now enters the exit site, and because it's no longer charged with an amino acid, the empty tRNA is released from the ribosome. The A-site is left empty, and the P-site is occupied by a tRNA holding its own amino acid and the amino acid of the preceding tRNA. The process of moving from one codon to the next is called *translocation* (not to be confused with the chromosomal translocations I describe in Chapter 15, where pieces of whole chromosomes are inappropriately swapped).

The ribosome continues to scoot along the mRNA in a 5' to 3' direction. The growing polypeptide chain is always attached to the tRNA that's sitting in the P-site, and the A-site is opened up repeatedly to accept the next charged tRNA. The process comes to a stop when the ribosome encounters one of the three stop codons. (For more on stop codons, see "Considering the combinations" earlier in this chapter.)

Termination

No tRNAs match the stop codon, so when the ribosome reads "stop," no more tRNAs enter the A-site (see Figure 10-6). At this point, a tRNA sits in the P-site with the newly constructed polypeptide chain attached to it by the tRNA's own amino acid. Special proteins called *release factors* move in and bind to the ribosome; one of the release factors recognizes the stop codon and sparks the reaction that cleaves the polypeptide chain from the last tRNA. After the polypeptide is released, the ribosome comes apart, releasing the final tRNA from the P-site. The ribosomal subunits are then free to find another mRNA to begin the translation process anew. Transfer RNAs are recharged with fresh amino acids and can be used over and over. Once freed, polypeptide chains assume their unique shapes and sometimes hook up with other polypeptides to carry out their jobs as fully functioning proteins (see the "Proteins Are Precious Polypeptides" section later in the chapter).

Messenger RNAs may be translated more than once and, in fact, may be translated by more than one ribosome at a time. As soon as the start codon emerges from the ribosome after the initiation of translation, another ribosome may recognize the mRNA's 5' cap, latch on, and start translating. Thus, many polypeptide chains can be manufactured very rapidly.

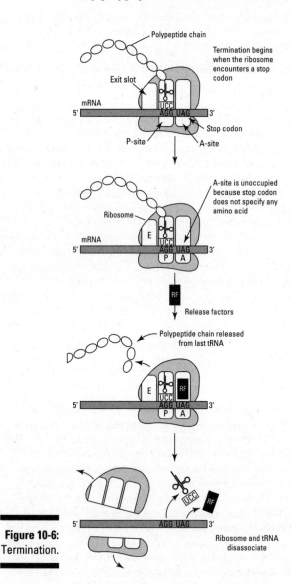

Figure 10-6: Termination.

Challenging the dogma

In other disciplines (say, physics), laws abound to describe the goings-on of the world. The law of gravity, for example, tolerates no violators. But genetics doesn't have laws because scientists keep acquiring new information. One exception is the *Central Dogma of Genetics.* A *dogma* isn't law; rather, it's more or less universally accepted opinion about how the world works. In this case, the Central Dogma of Genetics (coined by our old friend Francis Crick, of DNA-discovery fame; see Chapter 6) posited that the trip from genotype to phenotype is a one-way information highway.

Genetics seems to be a subject full of exceptions, and the Central Dogma is no . . . exception. Reverse transcription (that is, transmitting RNA's message back into DNA) does occur in some cases like viruses such as HIV, the virus that causes AIDS. RNA may also undergo replication (much as DNA does; see Chapter 7) transmitting information from RNA to RNA. Some evidence has even shown that DNA can be translated directly into protein without RNA at all (at least under laboratory conditions). In addition, there have been many new discoveries about the powerful roles that RNA has outside of translation. It turns out that many non-coding RNAs exist — that is, RNAs that don't code for proteins but play important roles in how genes are expressed (gene expression is the topic of Chapter 11).

Another idea that nearly attained the status of law was the *one gene–one polypeptide hypothesis. Polypeptides,* more familiarly known as proteins, are the products of gene messages. Back in the early 1940s, long before DNA was known to be the genetic material, two scientists, George Beadle and Edward Tatum, determined that genes code for proteins. Through a complex set of experiments, Beadle and Tatum discovered that each protein chain manufactured during translation is the product of only one gene's message. We now know that many different mRNA combinations are possible from a single gene. Each mRNA acts alone to make one polypeptide. Even here, there can be exceptions — some organisms may use a single codon to signal two different amino acids.

Proteins Are Precious Polypeptides

Besides water, the most common substance in your cells is protein. Proteins carry out the business of life. The key to a protein's function is its shape; completed proteins can be made of one or more polypeptide chains that are folded and hooked together. The way proteins fit and fold together depends on which amino acids are present in the polypeptide chains.

Recognizing radical groups

Every amino acid in a polypeptide chain shares several features, which you can see in Figure 10-7:

✔ A positively charged amino group (NH_2) attached to a central carbon atom

✔ A negatively charged carboxyl group (COOH) attached to the central carbon atom opposite the amino group

✔ A unique combination of atoms that form branches and rings, called *radical groups,* that differentiate the 20 amino acids specified by the genetic code

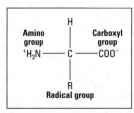

Radical groups

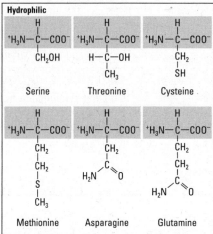

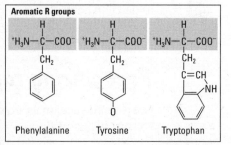

Figure 10-7: The 20 amino acids used to construct proteins.

Amino acid radical groups come in four flavors: water-loving (hydrophilic), water-hating (hydrophobic), negatively charged (bases), and positively charged (acids). When their amino acids are part of a polypeptide chain, radical groups of adjacent amino acids alternate sides along the chain (refer to Figure 10-7). Because of their differing affinities (those four flavors), the radical groups either repel or attract neighboring groups. This reaction leads to folding and gives each protein its shape.

Giving the protein its shape

Proteins are folded into complex and often beautiful shapes, as you can see in Figure 10-8. These arrangements are partly the result of spontaneous attractions between radical groups (see the preceding section for details) and partly the result of certain regions of polypeptide chains that naturally form spirals (also called *helices,* not to be confused with DNA's double helix in Chapter 6). The spirals may weave back and forth to form sheets. These spirals and sheets are referred to as a *secondary structure* (the simple, unfolded polypeptide chain is the *primary structure*).

Proteins are often modified after translation and may get hooked up with various other chemical groups and metals (such as iron). In a process similar to the post-transcription modification of mRNA, proteins may also be sliced and spliced. Some protein modifications result in natural folds, twists, and turns, but sometimes the protein needs help forming its correct conformation. That's what chaperones are for.

Chaperones are molecules that mold the protein into shape. Chaperones push and pull the protein chains until the appropriate radical groups are close enough to one another to form chemical bonds. This sort of folding is called a *tertiary structure.*

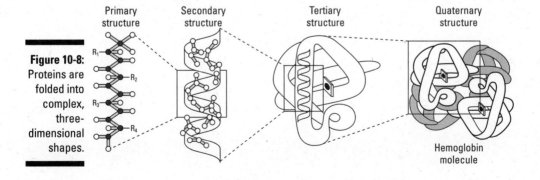

Figure 10-8: Proteins are folded into complex, three-dimensional shapes.

When two or more polypeptide chains are hooked to make a single protein, they're said to have a fourth degree, or *quaternary structure*. For example, the hemoglobin protein that carries oxygen in your blood is a well-studied protein with a quaternary structure. Two pairs of polypeptide chains form a single hemoglobin protein. The chains, two called *alpha-globin chains* and two called *beta-globin chains,* each form helices, which you can see in Figure 10-8, that wind around and fold back on themselves into tertiary structures. Associated with the tertiary structures are iron-rich *heme* groups that have a strong affinity for oxygen. For more on how good proteins go bad, flip to Chapters 11 and 13.

Chapter 11

Gene Expression: What a Cute Pair of Genes

*E*very cell in your body (with very few exceptions) carries the entire set of genetic instructions that make, well, everything about you. Your eye cells contain the genes for growing hair. Your nerve cells contain the genes that turn on cell division — yet your nerve cells don't divide (under normal conditions; see Chapter 14 for what happens when things go wrong). Genes that are supposed to be active in certain cells are turned on only when needed and then turned off again, like turning off the light in a room when you leave.

So why, then, aren't your eyeballs hairy? It boils down to gene expression. *Gene expression* is how genes make their products at the right time and in the right place. This chapter examines how your genes work and what controls them.

Getting Your Genes Under Control

Gene expression occurs throughout an organism's life, starting at the very beginning. When an organism develops — first as a *zygote* (the fertilized egg) and later as an embryo and fetus — genes turn on to regulate the process. At first, all the cells are exactly alike, but that characteristic quickly changes. (Cells that have the ability to turn into any kind of tissue are *totipotent*; see Chapter 20 for more on totipotency.) Cells get instructions from their DNA to turn into certain kinds of tissues, such as skin, heart, and bone. After the tissue type is decided, certain genes in each cell become active, and others get permanently turned off. That's because gene expression is highly *tissue-specific*, meaning certain genes are active only in certain tissues or at particular stages of development.

In part, the tissue-specific nature of gene expression is because of location — genes in cells respond to cues from the cells around them. Other than location, some genes respond to cues from the environment; other genes are set up to come on and then turn off at a certain stage of development. Take the genes that code for hemoglobin, for example.

Your *genome* (your complete set of genetic information) contains a large group of genes that all code for various components that make up the big protein, called *hemoglobin,* that carries oxygen in your blood. Hemoglobin is a complex structure comprised of two different types of proteins that are folded and joined together in pairs. During your development, nine different hemoglobin genes interacted at different times to make three kinds of hemoglobin. Changing conditions make it necessary for you to have three different sorts of hemoglobin at different stages of your life.

When you were still an embryo, your hemoglobin was composed mostly of epsilon-hemoglobin (Greek letters are used to identify the various types of hemoglobin). After about three months of development, the epsilon-hemoglobin gene was turned off in favor of two fetal hemoglobin genes (alpha and gamma). (*Fetal hemoglobin* is comprised of two proteins — two alphas and two gammas — folded and joined together as one functional piece.) When you were born, the gene producing the gamma-hemoglobin was shut off, and the beta-hemoglobin gene, which works for the rest of your life, kicked in.

Heat and light

Organisms have to respond quickly to changing conditions in order to survive. When external conditions turn on genes, it's called *induction.* Responses to heat and light are two types of induction that scientists understand particularly well.

When an organism is exposed to high temperatures, a suite of genes immediately kicks into action to produce *heat-shock proteins.* Heat has the nasty effect of mangling proteins so that they're unable to function properly, referred to as *denaturing.* Heat-shock proteins are produced by roughly 20 different genes and act to prevent other proteins from becoming denatured. Heat-shock proteins can also repair protein damage and refold proteins to bring them back to life. Heat-shock responses are best studied in fruit flies, but humans have a large number of heat-shock genes, too. These genes protect you from the effects of stress and pollutants.

Your daily rhythms of sleeping and waking are controlled, in part, by light. Even cancer may have a connection to light. When you're exposed to light during nighttime, your normal production of melatonin (a hormone that regulates sleep, among other things) is disrupted. In turn, a gene called *period* (so named because it controls circadian rhythms) is inactivated. Altered activity by the period gene is linked to breast cancer as well as depressed immune function. The increased incidence of breast cancer in women working the night shift was so dramatic that the researchers deemed night-shift work as a probable carcinogen.

The genes controlling the production of all these hemoglobins are on two chromosomes, 11 and 16 (see Figure 11-1). The genes on both chromosomes are turned on in order, starting at the 5' end of the group for embryonic hemoglobin (see Chapter 6 for how DNA is set up with numbered ends). Adult hemoglobin is produced by the last set of genes on the 3' end.

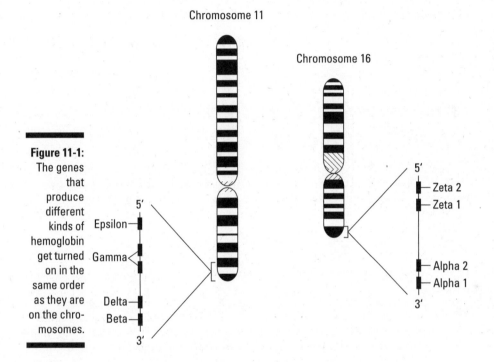

Figure 11-1: The genes that produce different kinds of hemoglobin get turned on in the same order as they are on the chromosomes.

Transcriptional Control of Gene Expression

Most gene control in eukaryotes, like you and me, occurs during transcription. I cover the basic transcription process in Chapter 9; this section covers how and when transcription is carried out to control when genes are and aren't expressed.

When a gene is "on," it's being transcribed. When the gene is "off," transcription is suspended. The only way that proteins (the stuff phenotype is made of; see Chapter 10) can be produced during translation is through the work of messenger RNA (mRNA). Transcription produces the mRNAs used in translation; therefore, when transcription is happening, translation is in motion, and gene expression is on. When transcription is stopped, gene expression is shut down, too. The timing of transcription can be controlled by a number of factors, including

✔ DNA accessibility

✔ Regulation from other genes

✔ Signals sent to genes from other cells by way of hormones

DNA must unwind a bit from its tight coils in order to be available for transcription to occur.

Tightly wound: The effect of DNA packaging

The default state of your genes is off, not on. Starting in the off position makes sense when you remember that almost every cell in your body contains a complete set of all your genes. You just can't have every gene in every cell flipped on and running amok all the time; you want specific genes acting only in the tissues where their actions are needed. Therefore, keeping genes turned off is every bit as important as turning them on.

Genes are kept in the off position in two ways:

✔ **Tight packaging:** DNA packaging is a highly effective mechanism to make sure that most genes are off most of the time because it prevents transcription from occurring by preventing transcription factors from getting access to the genes. DNA is an enormous molecule, and the only way it can be scrunched down small enough to fit into your cell's nuclei is by being tightly wound round and round itself in supercoils. First, the DNA is wrapped around special proteins called histones. Then, the DNA and the histones, which together look a bit like beads on a string, are wrapped around and around themselves to form the dense DNA known as *chromatin*. When DNA is wrapped up this way, it can't be transcribed, because transcription factors can't bind to the DNA to find the template strand and copy it. This is the heart of epigenetics (see Chapter 4).

✔ **Repressors:** *Repressors* are proteins that prevent transcription by binding to the same DNA sites that transcription activators would normally use or by interfering with the activities of the group of enzymes that kick off transcription (called the holoenzyme complex; see Chapter 9). In either case, DNA is prevented from unwinding, and the genes are kept turned off.

But genes can't stay off forever. Certain sections of DNA come prepackaged for unwinding, allowing the genes in those areas to be turned on more easily whenever they're needed.

To find out which genes are prepackaged for unwrapping, researchers exposed DNA to an enzyme called *DNase I,* which actually digests DNA. DNase I isn't a part of normal transcription; instead, it provides a signal to geneticists that a region of packaged DNA is less tightly wound than regions around it. Geneticists added DNase I to DNA to see which parts of the genome were sensitive to being degraded by the enzyme's activity. The sections of DNA left behind in these experiments contained genes that were always turned off in the tissue type the cell belonged to. The parts that were digested weren't tightly wound and thus harbored the genes that could be turned on when needed.

To turn genes on, the DNA must be removed from its packaging. To unwrap DNA from the nucleosomes, specific proteins must bind to the DNA to unwind it. Lots of proteins — including transcription factors, collectively known as *chromatin-remodeling complexes* — carry out the job of unwinding DNA depending on the needs of the organism. Most of these proteins attach to a region near the gene to be activated and push the histones aside to free up the DNA for transcription. As soon as the DNA is available, transcription factors, which in some types of cells are always lurking around, latch on and immediately get to work. As I explain in Chapter 9, transcription gets started when a group of enzymes called the *holoenzyme complex* binds to the promoter sequence of the DNA. Promoter sequences are part of the genes they control and are found a few bases away. *Transcription activator proteins* are part of the mix. These proteins help get all the right components in place at the gene at the right moment. Transcription activators also have the ability to shove histones out of the way to make the DNA template available for transcription.

Genes controlling genes

Four types of genes micromanage the activities of other genes. In this section, I divide these genes up into two groups based on how they relate to one another.

Micromanaging transcription

Three types of genes act as regulatory agents to turn transcription up *(enhancers),* turn it down *(silencers),* or drown out the effects of enhancing or silencing elements *(insulators).*

> ✔ **Enhancers:** This type of gene sequence turns on transcription and speeds it up, making transcription happen faster and more often. Enhancers can be upstream, downstream, or even smack in the middle of the transcription unit. Furthermore, enhancers have the unique ability to control genes that are distantly located (like thousands of bases away) from the enhancer's position. Nonetheless, enhancers are very tissue-specific in their activities — they only influence genes that are normally activated in that particular cell type.

Researchers are still working to get a handle on how enhancers do their jobs. Like the proteins that turn transcription on, enhancers seem to have the ability to rearrange nucleosomes and pave the way for transcription to occur. The enhancer teams up with transcription factors to form a complex called the *enhanceosome.* The enhanceosome attracts chromatin-remodeling proteins to the team along with RNA polymerase to allow the enhancer to supervise transcription directly.

✔ **Silencers:** These are gene sequences that hook up with repressor proteins to slow or stop transcription. Like enhancers, silencers can be many thousands of bases away from the genes they control. Silencers work to keep the DNA tightly packaged and unavailable for transcription.

✔ **Insulators:** Sometimes called *boundary elements,* these sequences have a slightly different job. Insulators work to protect some genes from the effects of silencers and enhancers, confining the activity of those sequences to the right sets of genes. Usually, this protection means that the insulator must be positioned between the enhancer (or silencer) and the genes that are off limits to the enhancer's (or silencer's) activities.

Given that enhancers and silencers are often far away from the genes they control, you may be wondering how they're able to do their jobs. Most geneticists think that the DNA must loop around to allow enhancers and silencers to come in close proximity to the genes they influence. Figure 11-2 illustrates this looping action. The promoter region begins with the TATA box and extends to the beginning of the gene itself. Enhancers interact with the promoter region to regulate transcription.

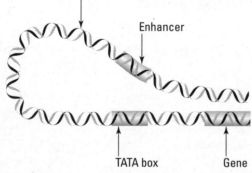

DNA

Enhancer

Figure 11-2:
Enhancers loop around to turn on genes under their control.

TATA box   Gene

Jumping genes: Transposable elements

Some genes like to travel. They hop around from place to place, inserting themselves into a variety of locations, causing mutations in genes, and changing the ways other genes do business. These wanderers are called *transposable elements* (TEs), and they're quite common — 50 percent of your DNA is made up of transposable elements, also known as *jumping genes.*

Barbara McClintock discovered TEs in 1948. She called them *controlling elements* because they control gene expression of other genes. McClintock was studying the genetics of corn when she realized that genes with a habit of frequently changing location were controlling kernel color. In her research, these genes showed up first on one chromosome, but in another individual, the genes mapped to a completely different chromosome. (You can find out more about Dr. McClintock in Chapter 22.)

It appears that TEs travel at will, showing up whenever and wherever they please. How they pull off this trick isn't completely clear, because TEs have several options when it comes to travel. They take advantage of breaks in DNA, but not just any break will do — the break must include little overhanging bits of single-stranded DNA (see Figure 11-3). Some TEs replicate themselves to hop into the broken spots. Others, which go by the special name *retrotransposons,* make use of RNA to do the job.

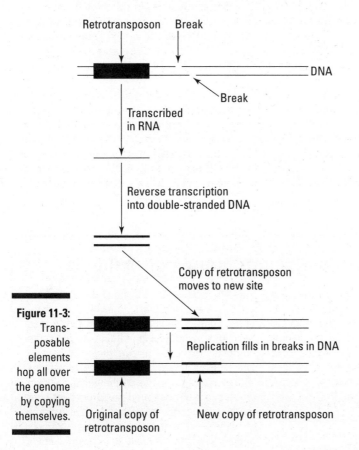

Figure 11-3: Transposable elements hop all over the genome by copying themselves.

Retrotransposons are transcribed just like all other DNA: An RNA transcript is produced. But then the RNA transcript is transcribed again by a special enzyme to make a double-stranded DNA copy of the RNA transcript. Because the result is a DNA copy made from an RNA transcript, the process used by retrotransposons is called *reverse transcription*. The DNA copy is then inserted into a break, and the newly copied retrotransposon makes itself comfortable.

Hormones turn genes on

Hormones are complex chemicals that control gene expression. They're secreted by a wide range of tissues in the brain, gonads (organs or glands, such as ovaries and testes, that produce reproductive cells), and other glands throughout the body. Hormones circulate in the bloodstream and can affect tissues far away from the hormones' production sites. In this way, they can affect genes in many different tissues simultaneously. Essentially, hormones act like a master switch for gene regulation all over the body. Take a look at the sidebar "Hormones make your genes go wild" for more about the effects hormones have on your body.

Some hormones are such large molecules that they often can't cross into the cells directly. These large hormone molecules rely on receptor proteins inside the cell to transmit their messages for them in a process called *signal transduction*. Other hormones, like steroids, are fat-soluble and small, so they easily pass directly into the cell to hook up with receptor proteins. Receptor proteins (and hormones small enough to enter the cell on their own) form a complex that moves into the cell nucleus to act as a transcription factor to turn specific genes on.

Hormones make your genes go wild

Dioxins are long-lived chemicals that are released into the environment through incineration of waste, coal-burning power plants, paper manufacturing, and metal smelting operations, to name a few. It turns out that dioxin can mimic estrogens and turn on genes all by itself. That's scary because it means that dioxin can cause cancer and birth defects.

Dioxin is a chemical with an unfortunate affinity for fat. Animals store dioxin in their fat cells, so most of the dioxins you're exposed to come in the food you eat. Meats and dairy products are the worst offenders, but fatty fish sometimes contain elevated levels, too. It's long been known that dioxins affect estrogens, the hormones that control reproduction in women and, to some degree, men, too. The good news is that dioxin levels are on the decline. Dioxin emissions have declined by 90 percent over the last 18 years. Unfortunately, dioxin that's already present in the environment breaks down slowly, so it's likely to persist for some time to come.

A swing and a miss: The genetic effects of anabolic steroids

Anabolic-androgenic steroids are in the news a lot these days. These steroids are synthetic forms of testosterone, the hormone that controls male sex determination (see Chapter 5). The anabolic aspect refers to chemicals that increase muscle mass; the androgenic aspect refers to chemicals that control gonad functions such as sex drive and, in the case of men, sperm production. High-profile athletes, including some famous baseball players, may have abused one or more of these drugs in an effort to improve performance. Reports also suggest that use of anabolic steroids is common among young athletes in high school and college.

Hormones like testosterone control gene expression. Research suggests that testosterone exerts its anabolic effects by depressing the activity of a tumor suppressor gene that produces the protein p27. When p27 is depressed in muscle tissue, the tissue's cells can divide more rapidly, resulting in the bulky physique prized by some athletes. Anabolic steroids apparently also accelerate the effects of the gene that causes male pattern baldness (see Chapter 5); thus, men carrying that allele and taking anabolic steroids become permanently bald faster and at a younger age than normal.

Defects in tumor-suppressor genes such as p27 are widely associated with cancer. Not only that, but some cancers depend on hormones to provide signals that tumor cells respond to (by multiplying). At least one study suggests that anabolic steroids are actually carcinogenic, meaning that their chemicals cause mutations that lead to cancer. Because illegally obtained steroids may also contain additional unwanted and potentially carcinogenic chemicals, mutagenic chemicals may be introduced into the body while simultaneously depressing the activity of a tumor-suppressor gene. It doesn't take a genius to realize that this is dangerous. Cancers associated with anabolic-androgenic steroid abuse include liver cancer, testicular cancer, leukemia, and prostate cancer.

The genes that react to hormone signals are controlled by DNA sequences called *hormone response elements* (HREs). HREs sit close to the genes they regulate and bind with the hormone-receptor complex. Several HREs can influence the same gene — in fact, the more HREs present, the faster transcription takes place in that particular gene.

Retroactive Control: Things That Happen after Transcription

After genes are transcribed into mRNA, their actions can still be controlled by events that occur later.

Interfering RNAs knock out genes

The world of RNAi (RNA interference; see "Shut up! mRNA silencing") is creating quite a splash in the understanding of how gene expression is controlled. The breakthrough moment came when two geneticists, Andrew Fire and Craig Mello, realized that by introducing certain double-stranded RNA molecules into roundworms, they could shut off genes at will. It turns out that scientists can put the RNAi into roundworm food and knock out gene function not only in the worm that eats the concoction but also in its offspring!

Since this discovery in 2003, geneticists have identified naturally occurring interfering RNAs in all sorts of organisms. The most well-known RNAi tend to be very short (only about 20 or so bases long) and hook up with special proteins, called *argonautes,* to regulate genes (mostly by silencing them). The argonaute proteins actually do the work, guided to the right target by the RNAi. RNAi finds its complementary mRNA (the product of the gene to be regulated), and the argonaute breaks down the mRNA, rendering it nonfunctional. New RNAi's are being discovered all the time, and their full importance in regulating genes is only just being realized. Longer, noncoding RNAs (over 200 bases long) are also produced during transcription; scientists are hard at work determining what functions those have.

The most promising applications for RNAi are in gene therapy (jump to Chapter 16 for that discussion). Using synthetic RNAi, geneticists have knocked out genes in all sorts of organisms, including chickens and mice. Work is also underway to knock out the function of genes in viruses and cancer cells.

Nip and tuck: RNA splicing

As you discover in Chapter 9, genes have sections called *exons* that actually code for protein products. Often, in between the exons are *introns,* interruptions of noncoding DNA that may or may not do anything. When genes are transcribed, the whole thing is copied into mRNA. The mRNA transcript has to be edited — meaning the introns are removed — in preparation for translation. When multiple introns are present in the unedited transcript, various combinations of exons can result from the editing process. Exons can be edited out, too, yielding new proteins when translation rolls around. This creative editing process allows genes to be expressed in new ways; one gene can code for more than one protein. This genetic flexibility is credited for the massive numbers of proteins you produce relative to the number of genes you have (see Chapter 9 for more on the potential of gene editing).

One gene in which genetic flexibility is very apparent is *DSCAM.* Named for the human disorder it's associated with — Down Syndrome Cell Adhesion Molecule — DSCAM may play a role in causing the mental disabilities that accompany Down syndrome. In fruit flies, DSCAM is a large gene with 115 exons and at least 100 splicing sites. Altogether, DSCAM is capable of coding for a whopping 30,016 different proteins. However, protein production from DSCAM is tightly regulated; some of its products only show up during early stages of fly development. The human version of DSCAM is less showy in that it makes only a few proteins, but other genes in the human genome are likely to be as productive at making proteins as DSCAM of fruit flies, making this a "fruitful" avenue of research. Humans have very few genes relative to the number of proteins we have in our bodies. Genes like DSCAM may help geneticists understand how a few genes can work to produce many proteins.

With scientists wise to the nip and tuck game played by mRNA, the next step in deciphering this sort of gene regulation is figuring out how the trick is done and what controls it. Researchers know that a complex of proteins called a *spliceosome* carries out much of the work in cutting and pasting genes together. How the spliceosome's activities are regulated is another matter altogether. Knowing how it all works will come in handy though, because some forms of cancer, most notably pancreatic cancer, can result from alternative splicing run amok.

Shut up! mRNA silencing

After transcription produces mRNA, genes may be regulated through *mRNA silencing*. mRNA silencing is basically interfering with the mRNA somehow so that it doesn't get translated. Scientists don't fully understand exactly how organisms like you and me use mRNA silencing, called *RNAi* (for *RNA interference*), to regulate genes. Geneticists know that most organisms use RNAi to stymie translation of unwanted mRNAs and that double-stranded RNA provides the signal for the initiation of RNAi, but the details are still a mystery. The discovery of RNAi has produced a revolution in the study of gene expression; see the sidebar "Interfering RNAs knock out genes" for more.

RNA silencing isn't just used to regulate the genes of an organism; sometimes it's used to protect an organism from the genes of viruses. When the organism's defenses detect a double-stranded virus RNA, an enzyme called *dicer* is produced. Dicer chops the double-stranded RNA into short bits (about 20 or 25 bases long). These short strands of RNA, now called *small interfering RNAs* (siRNAs), are then used as weapons against remaining viral RNAs. The siRNAs turn traitor, first pairing up with RNA-protein complexes produced by the host and then guiding those complexes to intact viral RNA. The viral RNAs are then summarily destroyed and degraded.

mRNA expiration dates

After mRNAs are sliced, diced, capped, and tailed (see Chapter 9 for how mRNA gets dressed up), they're transported to the cell's cytoplasm. From that moment onward, mRNA is on a path to destruction because enzymes in the cytoplasm routinely chew up mRNAs as soon as they arrive. Thus, mRNAs have a relatively short lifespan, the length of which (and therefore the number of times mRNA can be translated into protein) is controlled by a number of factors. But the mRNA's poly-A tail (the long string of adenines tacked on to the 3' end) seems to be one of the most important features in controlling how long mRNA lasts. Key aspects of the poly-A tail include:

✓ **Tail length:** The longer the tail, the more rounds of translation an mRNA can support. If a gene needs to be shut off rapidly, the poly-A tail is usually pretty short. With a short tail, when transcription comes to a halt, all the mRNA in the cytoplasm is quickly used up without replacement, thus halting protein production, too.

✓ **Untranslated sequences before the tail:** Many mRNAs with very short lives have sequences right before the poly-A tail that, even though they aren't translated, shorten the mRNA's lifespan.

Hormones present in the cell may also affect how quickly mRNAs disappear. In any event, the variation in mRNA expiration dates is enormous. Some mRNAs last a few minutes, meaning those genes are tightly regulated; other mRNAs hang around for months at a time.

Gene Control Lost in Translation

Translation of mRNA into amino acids is a critical step in gene expression. (Flip to Chapter 10 for a review of the players and process of translation.) But sometimes genes are regulated during or even after translation.

Modifying where translation occurs

One way gene regulation is enforced is by hemming in mRNAs in certain parts of the cytoplasm. That way, proteins produced by translation are found only in certain parts of the cell, limiting their utility. Embryos use this strategy to direct their own development. Proteins are produced on different sides of the egg to create the front and back, so to speak, of the embryo.

Modifying when translation occurs

Just because an mRNA gets to the cytoplasm doesn't mean it automatically gets translated. Some gene expression is limited by certain conditions that block translation from occurring. For example, an unfertilized egg contains lots of mRNAs supplied by the female. Translation actually occurs in the unfertilized egg, but it's slow and selective. All that changes when a sperm comes along and fertilizes the egg: Preexisting mRNAs are slurped up by waiting ribosomes, which are signaled by the process of fertilization. New proteins are then rapidly produced from the maternal mRNAs.

Controlling gene expression by controlling translation occurs in one of two ways:

 ✔ The machinery that carries out translation, such as the initiator proteins that interact with ribosomes, is modified to increase or decrease how effectively translation occurs.

 ✔ mRNA carries a message that controls when and how it gets translated.

All mRNAs carry short sequences on their 5' ends that aren't translated, and these sequences can carry messages about the timing of translation. The untranslated sequences are recognized with the help of translation initiation factors that help assemble the ribosome at the start codon of the mRNA. Some cells produce mRNAs but delay translation until certain conditions are met. Some cells respond to levels of chemicals that the cell's exposed to. For example, the protein that binds to iron in the blood is created by translation only when iron is available, even though the mRNAs are being produced all the time. In other cases, the condition of the organism sends the message that controls the timing of translation. For example, insulin, the hormone that regulates blood sugar levels, controls translation, but when insulin's absent, the translation factors lock up the needed mRNAs and block translation from occurring. When insulin arrives on the scene, the translation factors release the mRNAs, and translation rolls on, unimpeded.

Modifying the protein shape

The proteins produced by translation are the ultimate form of gene expression. Protein function, and thus gene expression, can be modified in two ways: by changing the protein's shape or by adding components to the protein. The products of translation, the amino acid chains, can be folded in various ways to affect their functions (see Chapter 9 for how amino acid chains are folded). Various components — carbohydrate chains, phosphates, and metals such as iron — can be added to the chain, also changing its function. Occasionally, the folding of proteins can go horribly wrong; for an explanation of one of the scariest products of this type of error, mad cow disease, check out the sidebar "Proteins gone wrong."

Proteins gone wrong

Cruetzfeldt-Jakob disease (CJD) is a frightening disorder of the brain. Sufferers first experience memory loss and anxiety, and they ultimately develop tremors and lose intellectual function. CJD is the human form of what's popularly known as *mad cow disease*. The pathogen isn't a bacteria, virus, or parasite — it's an infectious protein called a *prion*. One of the scariest aspects of prions is that they seem to be able to replicate on their own by hijacking normal proteins and refolding them.

The gene that codes for the prion protein is found in many different organisms, including humans. After it's mutated (and what the unmutated version does isn't really clear), the protein produced by the gene folds into an unusual, flattened sheet. After one prion protein is acquired, that prion can hijack the normal products of unmutated prion genes, turning them into misfolded monsters, too. Prion proteins gum up the brain of the affected organism and eventually have fatal results. As if this outcome weren't frightening enough, it seems that prions can jump from one species to another.

Scientists are fairly certain that some of the cows originally infected by mad cow disease contracted it by eating feed contaminated by sheep meat. The deceased sheep were infected with a prion that causes yet another icky disease called *scrapie,* which destroys the brains of infected animals. Scientists believe that when humans consume beef products from cows affected by mad cow disease, the prions in the meat can migrate through the human body and continue doing their dirty work.

Part III
Genetics and Your Health

The 5th Wave   By Rich Tennant

"You can do all the DNA testing you want Pinocchio, but I still feel this is your baby."

In this part . . .

Genetics affects your everyday life. Viruses, bacteria, parasites, and hereditary diseases all have their roots in DNA. That's why as soon as scientists uncovered the chemical nature of DNA, the race was on to read the code directly.

Genetic information is used to track, diagnose, and treat genetic diseases. The chapters in this part help you unravel the mysterious connections between DNA and your health. I explain how genetic counselors read your family tree to help you better understand your family medical history. I cover the ways in which mutations alter genes and the consequences of those changes. And because serious problems arise when chromosomes aren't doled out in the usual way — leading to too many or too few — I explain what the numbers mean. Finally, I share some exciting information about how genetics may someday reshape medical treatments in the form of gene therapies.

Chapter 12

Genetic Counseling

*1*f you're thinking of starting a family or adding to your brood, you may be wondering what your little ones will look like. Will they get your eyes or your dad's hairline? If you know your family's medical history, you may also have significant worries about diseases such as cystic fibrosis, Tay-Sachs, or sickle cell anemia. You may worry about your own health, too, as you contemplate news stories dealing with cancer, heart disease, and diabetes, for example. All these concerns revolve around genetics and the inheritance of a predisposition for a particular disease or the inheritance of the disorder itself.

Genetic counselors are specially and rigorously trained to help people learn about the genetic aspects of their family medical histories. This chapter explains the process of genetic counseling, including how counselors generate family trees and estimate probability of inheritance and how genetic testing is done when genetic disorders are anticipated.

Getting to Know Genetic Counselors

Like it or not, you have a family. You have a mother and a father, grandparents, and perhaps children of your own. You may not think of them, but you also have hundreds of ancestors — people you've never met — whose genes you carry and may pass down to descendants in the centuries to come.

Genetic counselors help people like you and me examine our families' genetic histories and uncover inherited conditions. They work with medical personnel like physicians and nurses to interpret medical histories of patients and their families. Although they aren't trained as geneticists, they usually hold a master's degree in genetic counseling and have an extensive background in genetics (and can solve genetics problems in a snap; see

Chapters 3 through 5 for some examples) so that they can spot patterns that signal an inherited disorder. (For more on genetic counselors and other career paths in genetics, see Chapter 1.)

Genetics counselors perform a number of functions, including

- Constructing and interpreting family trees, sometimes called *pedigrees,* to assess the likelihood that various inherited conditions will be (or have been) passed on to a particular generation.

- Counseling families about options for diagnosis and treatment of genetic conditions.

Physicians most commonly refer the following types of people or patients to genetic counselors:

- Couples who are concerned about exposure to substances known to cause birth defects (such as radiation, viruses, drugs, and chemicals)

- Couples who have experienced more than one miscarriage or stillbirth or who have problems with infertility

- Parents of a child who shows symptoms of a genetic disorder

- People with a family history of a particular disorder, such as cystic fibrosis, who are planning a family

- People with a family history of inherited diseases like Parkinson disease or certain cancers such as breast, ovarian, or colon cancer who may be considering genetic testing to determine their risk of getting the disease

- Women over 35 who are pregnant or planning a pregnancy

- Women who have had an abnormal screening test, such as an ultrasound, during a pregnancy

I cover many of the scientific reasons for the inheritance of genetic disorders elsewhere in this book. Mutations within genes are the root cause of many genetic disorders (including cystic fibrosis, Tay-Sachs disease, and sickle cell anemia), and I cover mutation in detail in Chapter 13. I discuss the causes and genetic mechanics of cancer in Chapter 14. I explain chromosomal disorders such as Down syndrome, trisomy 13, and fragile X syndrome in Chapter 15. Finally, I cover gene therapy treatments for inherited disorders in Chapter 16.

Building and Analyzing a Family Tree

Often the first step in genetic counseling is drawing a family tree. The tree usually starts with the person for whom the tree is initiated; this person is called the *proband.* The proband can be a newly diagnosed child, a woman planning a pregnancy, or an otherwise healthy person who's curious about risk for inherited disease. Often, the proband is simply the person who meets with the genetic counselor and provides the information used to plot out the family tree. The proband's position in the family tree is always indicated by an arrow, and he or she may or may not be affected by an inherited disorder.

Genetic counselors use a variety of symbols on family trees to indicate personal traits and characteristics. For instance, certain symbols convey sex, gene carriers, whether the person is deceased, and whether the person's family history is unknown. The manner in which symbols are connected show relationships among people, such as which offspring belong to which parents, whether someone is adopted, and whether someone is a twin. Check out Figure 12-1 for a detailed key to the symbols typically used in pedigree analysis.

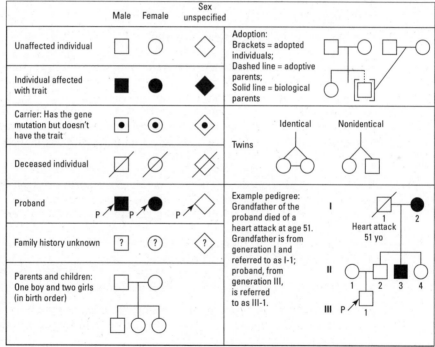

Figure 12-1:
Symbols commonly used in pedigree analysis.

In a typical pedigree, the age or date of birth of each person is noted on the tree. If deceased, the person's age at the time of death and the cause of death are listed. Some genetic traits are more common in certain regions of the world, so it's useful to include all kinds of other details about family history on the pedigree, such as what countries people immigrated from or how they're related. Every member of the family should be listed, along with any medical information known about that person, including the age at which certain medical disorders occurred. In the example in Figure 12-1, the grandfather of the proband died of a heart attack at age 51. Including this information creates a record of all disorders with the relation to the family tree so that the counselor is more likely to detect every inherited disease present in the family. (Medical information doesn't appear in Figure 12-1, but it's normally a part of a tree.)

Medical problems often listed on pedigrees include

- Alcoholism or drug addiction
- Asthma
- Birth defects, miscarriages, or stillbirths
- Cancer
- Heart disease, high blood pressure, or stroke
- Kidney disease
- Mental illness or mental retardation

Human couples have only a few children relative to other creatures, and humans start producing offspring after a rather long childhood. Geneticists rarely see neat offspring ratios (such as four siblings with three affected and one unaffected) in humans that correspond to those observed in animals (take a look at Chapters 3 and 4 for more on common offspring ratios). Therefore, genetic counselors must look for very subtle signs to detect particular patterns of inheritance in humans.

When the genetic counselor knows what kind of disorder or trait is involved, he or she can determine the likelihood a particular person will possess the trait or pass it on to his or her children. (Sometimes, the disorder is unidentified, such as when a person has a family history of "heart trouble" but doesn't have a precise diagnosis.) Genetic counselors use the following terms to describe the individuals in a pedigree:

- **Affected:** Any person having a given disorder.
- **Heterozygote:** Any person possessing one copy of the mutated gene coding for a disorder (an allele; see Chapter 2 for details). An unaffected heterozygote is called a *carrier.*
- **Homozygote:** Any person possessing two copies of the allele for a disorder. This person can also be described as *homozygous.*

The particular way in which most human genetic disorders are passed down to later generations — the *mode of inheritance* — is well established. After a genetic counselor determines which family members are affected or are likely to be carriers, it's relatively easy to determine the probability of another person being a carrier or inheriting the disorder.

In the following sections, I explore the modes of inheritance for human genetic disorders, how genetic counselors map these modes, and how you (and your counselor) can figure out the probability of passing these traits on to offspring. For additional background on each of these modes of inheritance and the subject of inheritance in general, see Chapters 3 through 5.

Autosomal dominant traits

A *dominant* trait or disorder is one that's expressed (or manifested) in anyone who inherits the mutation for the trait. *Autosomal dominant* means that the gene is carried on a chromosome other than a sex chromosome (meaning not on an X or a Y; see Chapter 3 for more details). In human pedigrees, autosomal dominant traits have some typical characteristics:

✔ Affected children are born to an affected parent.

✔ Both males and females are affected with equal frequency.

✔ If neither parent is affected, usually no child is affected.

✔ The trait doesn't skip generations.

Figure 12-2 shows the pedigree of a family with an autosomal dominant trait. In the figure, affected persons are shaded, and you can clearly see how only affected parents have affected children. The trait can be passed to a child from either the mother or the father. Generally, affected parents have a 50-percent chance of passing an autosomal dominant trait or disorder on to each child.

Some common autosomal dominant disorders are

✔ Achondroplasia, a form of dwarfism

✔ Huntington disease, a progressive and fatal disease affecting the brain and nervous system

✔ Marfan syndrome, a disorder affecting the skeletal system, heart, and eyes

✔ Polydactyly, or extra fingers and toes

The normal pattern of autosomal dominant inheritance has three exceptions:

✔ **Reduced penetrance:** *Penetrance* is the percentage of individuals having a particular gene (genotype) that actually display the physical characteristics dictated by the gene (or express the gene as phenotype, scientifically speaking; see Chapter 3 for a full rundown of genetics terms). Many autosomal dominant traits have complete penetrance, meaning that every person inheriting the gene shows the trait. But some traits have *reduced penetrance,* meaning only a certain percentage of individuals inheriting the gene show the phenotype.

When an autosomal dominant disorder shows reduced penetrance, the phenotype skips generations. Check out Chapter 3 for more details on reduced penetrance.

✔ **New mutations:** In the case of new mutations that are autosomal dominant, the trait appears for the first time in a particular generation and can appear in every generation thereafter. You can flip to Chapter 13 to find out more details about mutations — how they occur and how they're passed on.

✔ **Variable expressivity:** Expressivity is the degree to which a trait is expressed. Some conditions may be undiagnosed in earlier generations because the condition is so mild, it goes undetected. Turn to Chapter 4 to find out more about expressivity.

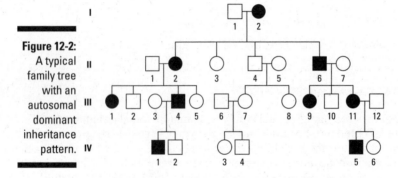

Figure 12-2: A typical family tree with an autosomal dominant inheritance pattern.

Autosomal recessive traits

Recessive disorders are expressed only when an individual inherits two identically altered (or mutated) copies of the gene that causes the disorder. It's then said that the individual is *homozygous* for that gene (see Chapter 3 for more details on inheritance). Like autosomal dominant disorders, autosomal recessive disorders are coded in genes found on chromosomes other than sex chromosomes. In pedigrees, such as the one in Figure 12-3, autosomal recessive disorders typically have the following characteristics:

✔ Affected children are born to unaffected parents.

✔ Both males and females are affected equally.

✔ Children born to parents who share common ancestry (such as ethnic or religious background) are more likely to be affected than those of parents with different backgrounds.

✔ The disorder or trait skips one or more generations, or is present only in a single generation (siblings).

The probability of inheriting an autosomal recessive disorder varies depending on which alleles parents carry (see Chapter 3 for all the details on how the odds of inheritance are calculated):

- ✔ **When both parents are carriers,** every child born to the couple has a 25 percent chance of being affected.

- ✔ **When one parent is a carrier and the other isn't,** every child has a 50 percent chance of being a carrier. No child will be affected.

- ✔ **When one parent is a carrier and the other is affected,** each child has a 50 percent chance of being affected. All unaffected children from the union will be carriers.

- ✔ **When one parent is affected and the other is unaffected (and not a carrier),** all children born to the couple will be carriers. No children will be affected.

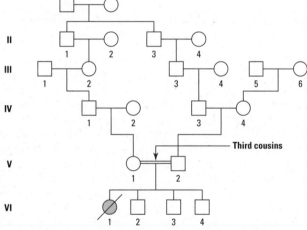

Figure 12-3:
A typical autosomal recessive disorder in a family tree.

Cystic fibrosis (CF) is an autosomal recessive disorder that causes severe lung and digestive problems in affected persons. As with all autosomal recessive disorders, if both members of a couple are carriers for cystic fibrosis, they have a 25 percent chance of having an affected child with each pregnancy they have. That's because both the man and the woman are heterozygous for the allele that codes for cystic fibrosis, and each has a 50 percent probability of contributing the CF allele. You calculate the probability of *both* members of the couple contributing CF alleles in one fertilization event by multiplying the probability of each event happening independently. The probability the father contributes his CF allele is 50 percent, or 0.5; the probability the mother contributes her CF allele is also 50 percent, or 0.5. The probability that both contribute their allele is $0.5 \times 0.5 = 0.25$, or 25 percent. For more details on how to calculate probabilities of inheritance, flip to Chapters 3 and 4.

Genetic disorders in small populations

The Pennsylvania Amish don't have electricity in their homes, don't drive cars, and don't use e-mail or cellphones. They live simply in the modern world as a religious way of life. Because Amish people marry within their faith, certain genetic disorders are common. Amish families come by horse and buggy to the Clinic for Special Children in Strasburg, Pennsylvania, for genetic testing. By partnering with an ultra–high-tech company, the clinic provides rapid, inexpensive genetic testing. Among the clinic's findings is the fact that the Old Order Amish of southeastern Pennsylvania suffer from a devastating form of sudden infant death syndrome (SIDS). Altogether, the Belleville Amish community has mourned the loss of over 21 babies (one family lost six infants to the disorder). Researchers at the Translational Genomics Research Institute in Phoenix, Arizona, were able to locate the mutated gene that causes the SIDS using microarray technology (see Chapter 23). Sadly, no treatment yet exists for this type of SIDS, but gene therapy (which I cover in Chapter 16) may offer hope for small populations such as the Amish.

Some autosomal recessive disorders are more common among people of certain religious or ethnic groups, because people belonging to those groups tend to marry within the group. After many generations, everyone within the group shares common ancestry. When cousins or other close relatives marry, such relationships are referred to as *consanguineous* (meaning "same blood"). Generally, people who are more distantly related than fourth cousins aren't considered "related," but in fact, those persons still share alleles from a common ancestor. When populations are founded by rather small groups of people, those groups often have higher rates of particular genetic disorders than the general population; for more details, take a look at the sidebar "Genetic disorders in small populations." In these cases, autosomal recessive disorders may no longer skip generations, because so many persons are heterozygous and thus carriers of the disorder.

X-linked recessive traits

Males are XY and therefore have only one copy of the X chromosome; they don't have a second X to offset the expression of a mutant allele on the affected X. Thus, similar to autosomal dominant disorders, X-linked recessive disorders express the trait fully in males, even though they're not homozygous. Females rarely show X-linked recessive disorders, because being homozygous for the disorder is very rare. In pedigrees, X-linked recessive disorders have the following characteristics:

✓ Affected sons are born to unaffected mothers.

✓ Far more males than females are affected.

✓ The trait is *never* passed from father to son.

✓ The disorder skips one or more generations.

Unaffected parents can have unaffected daughters and one or more affected sons. Women who are carriers frequently have brothers with the disease, but if families are small, a carrier may have no affected immediate family members. Sons of affected fathers are never affected, but daughters of affected fathers are always carriers, because daughters must inherit one of their X chromosomes from their fathers. In this case, that X chromosome will always carry the allele for the disorder. The pedigree in Figure 12-4 is a classic example of a well-researched family possessing many carriers for the X-linked disorder hemophilia, a devastating disorder that prevents normal clotting of the blood. For more on the royal families whose history is pictured in Figure 12-4, see the sidebar "A royal pain in the genes."

The probability of inheritance of X-linked disorders depends on gender. Female carriers have a 50 percent likelihood of passing the gene on to each child. Males determine the gender of their offspring, making the chance of any particular child being a boy 50 percent. Therefore, the likelihood of a carrier mom having an affected son is 25 percent (chance of having a son = 0.5; chance of a son inheriting the affected X = 0.5; therefore, $0.5 \times 0.5 = 0.25$, or 25 percent).

Figure 12-4: The X-linked recessive disorder hemophilia works its way through the pedigree of the royal families of Europe and Russia.

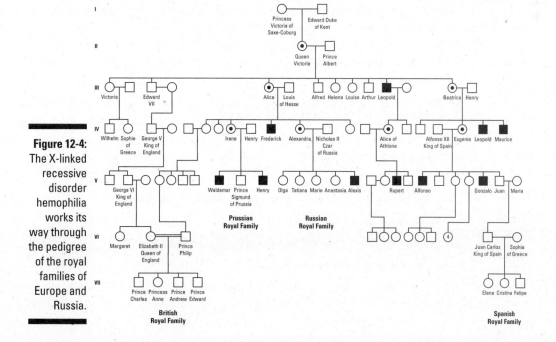

A royal pain in the genes

You can find one of the most famous examples of an X-linked family pedigree in the royal families of Europe and Russia, which you can see in Figure 12-4. Queen Victoria of England had one son affected with hemophilia. It's not clear whom Queen Victoria inherited the allele from; she may have been the victim of spontaneous gene mutation. In any event, two of her daughters were carriers, and she had one affected son, Leopold. Queen Victoria's granddaughter Alexandra was also a carrier. Alexandra married Nicholas Romanov, who became czar of Russia, and together they had five children: four daughters and one son. The son, Alexis, suffered from hemophilia.

The role Alexis's disease played in his family's ultimate fate is debatable. Clearly, however, one of the men who influenced the downfall of Russia's royal family was linked to the family as Alexis's "doctor." Gregory Rasputin was a self-proclaimed faith healer; in photographs, he appears wild-eyed and deeply intense. He's generally perceived to have been a fraud, but at the time, he had a reputation for miraculous

healings, including helping little Alexis recover from a bleeding crisis. Despite Rasputin's talent for healing, Alexis didn't live to see adulthood. Shortly after the Russian Revolution broke out, the entire Russian royal family was murdered. (Rasputin himself had been murdered some two years earlier.)

In a bizarre final twist to the Romanov tale, a road repair crew discovered the family's bodies in 1979. Oddly, two of the family members were missing. Eleven people were supposedly killed by firing squad on the night of July 16, 1918: the Russian royal family (Alexandra, Nicholas, and their five children) along with three servants and the family doctor. However, the bodies of Alexis and his little sister, Anastasia, have never been found. Using DNA fingerprinting, researchers confirmed the identities of Alexandra and her children by matching their mitochondrial DNA to that of one of Queen Victoria's living descendants, Prince Philip of England. (To find out more about the forensic uses of DNA, flip to Chapter 18.)

X-linked dominant traits

Like autosomal dominant disorders, X-linked dominant traits don't skip generations. Every person who inherits the allele expresses the disorder. The family tree in Figure 12-5 shows many of the hallmarks of X-linked dominant disorders:

✔ Affected mothers have both affected sons and daughters.

✔ Both males and females are affected.

✔ All daughters of affected fathers are affected.

✔ The trait doesn't skip generations.

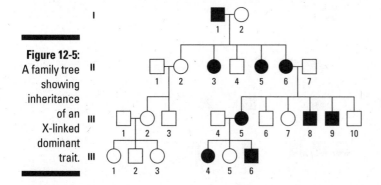

Figure 12-5:
A family tree
showing
inheritance
of an
X-linked
dominant
trait.

X-linked dominant traits show up more often in females than males because females can inherit an affected X from either parent. In addition, some disorders are lethal in males who are _hemizygous_ (having only one copy of the chromosome, not two; see Chapter 5). Affected females have a 50 percent chance of having an affected child of either sex. Males never pass their affected X to sons; therefore, sons of affected fathers and unaffected mothers have _no_ chance of being affected, in contrast to daughters, who are always affected. The probability of an affected man having an affected child is 50 percent (that is, equal to the likelihood of having a daughter).

Y-linked traits

The Y chromosome is passed strictly from father to son. By definition, Y-chromosome traits are considered hemizygous. Y-chromosome traits are expressed as if they were dominant because there's only one copy of the allele per male, with no other allele to offset the effect of the gene. Y-linked traits are easy to recognize when seen in a pedigree, such as Figure 12-6, because they have the following characteristics:

✔ Affected men pass the trait to all their sons.

✔ No women are ever affected.

✔ The trait doesn't skip generations.

Because the Y chromosome is tiny and has relatively few genes, Y-linked traits are very rare. Most of the genes involved control male-only traits such as sperm production and testis formation. If you're female and your dad has hairy ears, you can relax — hairy ears is also considered a Y-linked trait.

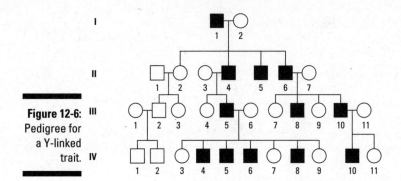

Figure 12-6:
Pedigree for
a Y-linked
trait.

Genetic Testing for Advance Notice

With the advent of many new technologies (some of which grew out of the Human Genome Project, which I explain in Chapter 8), genetic testing is easier and cheaper than ever. Genetic testing and genetic counseling often go hand in hand. The genetic counselor works to identify which disorders occur in the family, and testing then examines the DNA directly to determine whether the disorder-causing gene is present. Your physician may refer you or a family member for genetic testing for a variety of reasons, particularly if you

- Are a healthy person concerned about certain heritable disorders in your ethnic background or family such as breast cancer or Huntington disease

- Are a healthy person with a family history of a recessive disorder, and you're thinking about having a child

- Are a pregnant woman over 35

- Are an affected person and need to confirm a diagnosis

- Have an infant who's at risk (because his or her parents are known or suspected carriers)

General testing

Every person the world over carries one or more alleles that cause genetic disease. Most of us never know which alleles or how many we carry. If you have a family member who's affected with a rare genetic disorder, particularly an autosomal dominant disorder with incomplete penetrance or delayed onset, you may be vitally concerned about which allele(s) you carry. Persons currently unaffected with certain disorders can seek genetic testing to learn if they're carriers. Most tests involve a blood sample, but some are done with a simple cheek swab to capture a few skin cells. You can find more about genetic testing for inherited disorders in Chapter 13 and about testing for inherited forms of cancer in Chapter 14. Genetic testing has many ethical implications, as I cover in Chapter 21.

Prenatal testing

Prenatal diagnosis is commonly used for unborn children of women over 35, because such women are much more likely than younger women to have children with chromosomal disorders (see Chapter 15). Prenatal testing is designed to allow time for couples to make decisions about treatments to be administered either during pregnancy or after delivery of an affected infant.

Chorionic villus sampling and amniocentesis

For definitive diagnosis of a genetic disorder, testing requires tissue of the affected person. Two common prenatal tests used to obtain fetal tissue for testing are *chorionic villus sampling* (CVS) and *amniocentesis*. Both tests require ultrasound to guide the instruments used to obtain the samples (see the following section for more info on ultrasound).

- ✔ **CVS** is usually done late in the first trimester of pregnancy (weeks 10 to 12). A catheter is inserted vaginally and guided to the outer layer of the placenta, called the *chorion.* Gentle suction is used to collect a small sample of chorionic tissue. The placental tissue arises from the fetus, not the mother, so the collected cells give an accurate picture of the fetus's chromosome number and genetic profile. The advantages of CVS are that it can be done earlier than most other prenatal genetic tests; it's extremely accurate; and because a relatively large sample is obtained, results are rapidly produced. CVS is associated with a slightly higher rate of miscarriage.

- ✔ **Amniocentesis** is usually done early in the second trimester of pregnancy (weeks 15 and beyond). Amniocentesis is used to obtain a sample of the amniotic fluid that surrounds the growing fetus, because amniotic fluid contains fetal cells (skin cells that have sloughed off) that can be examined for prenatal testing. The fluid is drawn directly from the uterus using a needle inserted through the abdomen. Because fetal cells in the fluid are at a very low concentration, the cells must be grown in a lab to provide enough tissue for testing, making results slow to come (about one to two weeks). But the results are accurate, and complications following the procedure (such as miscarriage) are rare.

Ultrasound

Ultrasound technology allows physicians to examine a growing fetus visually, along with its spinal cord, brain, and all its organs. Ultrasound can be done much earlier in a woman's pregnancy than CVS or amniocentesis.

Ultrasound directs extremely high frequency sound waves through the mother's abdominal wall. The sound waves bounce off the fetus and return to a receiver that then converts the sound wave "picture" into a visual image. New ultrasound technologies include powerful computers that put together a three-dimensional image, giving amazingly crisp pictures of facial features and body parts. Ultrasound is generally used to screen for genetic disorders associated with physical features or deformities. Ultrasound can be used at any time during pregnancy and is completely non-invasive, with little or no risk to mother or baby.

Newborn screening

Some genetic disorders are highly treatable using dietary restrictions. Therefore, all newborns in the United States are tested for two common, highly treatable genetic disorders: *phenylketonuria* and *galactosemia*. Both of these disorders are autosomal recessive.

- **Phenylketonuria** causes mental retardation due to the buildup of *phenylalanine* (an amino acid that's part of a normal diet) in the brain of affected persons. A diet low in phenylalanine allows such persons to live symptom-free lives. (This disorder and the potential to control it are the reasons certain diet colas contain warning labels regarding phenylalanine content.) Phenylketonuria occurs once in every 10,000 to 20,000 births.

- **Galactosemia** is a disorder similar to phenylketonuria that results from an inability to break down one of the products of lactose (milk sugar). A lactose-free diet allows affected persons to live symptom-free lives. If untreated, galactosemia results in brain damage, kidney and liver failure, and often death. Galactosemia occurs once in every 45,000 births.

Testing for these two disorders isn't actually genetic testing; rather, the tests are designed to look for the presence of abnormal amounts of either phenylalanine or galactose — the phenotypes of the disorders. As technologies advance, these tests may be replaced with direct DNA examination by gene chips (which you can read more about in Chapter 23).

Chapter 13

Mutation and Inherited Diseases: Things You Can't Change

Despite what you may think, mutation is good thing. *Mutation,* which is simply genetic change, is responsible for all phenotypic variation. Variation in flower colors and plant height, the flavor of different varieties of apples, the differences among dog breeds, you name it — the natural process that created all those different phenotypes is mutation. Mutation occurs all the time, spontaneously and pretty much randomly.

But like many good things, mutation can also be bad. It can disrupt normal gene activity and cause disease such as cancer (flip to Chapter 14 for details) and birth defects (see Chapter 15). In this chapter, you discover what causes mutations, how DNA can repair itself in the face of mutation, and what the consequences are when repair attempts fail.

Sorting Out Types of Mutations

Mutations fall into two major categories, and the distinction between the two is important to keep in mind:

> ✔ **Somatic mutations:** Mutations in body cells that don't make eggs or sperm. Mutations that occur in the somatic cells aren't *heritable* — that is, the changes can't be passed from parent to offspring — but they do affect the person with the mutation.

✔ **Germ-cell mutations:** Mutations in the sex cells (germ cells like eggs and sperm; see Chapter 2 for the scoop on cell types) that lead to embryo formation. Unlike somatic mutations, germ-cell mutations often don't affect the parent. Instead, they affect the offspring of the person with the mutation and are heritable from then on.

Some disorders have elements of both somatic and germ-cell (heritable) mutations. Many cancers that run in families arise as a result of somatic mutations in persons who are already susceptible to the disease because of mutations they inherited from one or both parents. (You can find out more about heritable cancers in Chapter 14.)

Both somatic and germ-cell mutations usually come about, in a general sense, because of

✔ **Substitutions of one base for another:** Substitutions are sometimes called *point mutations*. Usually, only one mistaken base is involved, although sometimes both the base and its complement are changed (for a review of the chemistry of DNA, turn to Chapter 6). This type of mutation breaks down further into two categories:

  • **Transition mutation:** When a purine base is substituted for the other purine, or one pyrimidine is substituted for the other pyrimidine. Transition mutations are the most common form of substitution errors.

  • **Transversion mutation:** When a purine replaces a pyrimidine (or vice versa).

✔ **Insertions and deletions of one or more bases:** When an extra base is added to a strand, the error is called an *insertion*. Dropping a base is considered a *deletion*. Insertions and deletions are the most common forms of mutation. When the change happens within a gene, both insertions and deletions lead to a change in the way the genetic code is read during translation (flip to Chapter 10 for a translation review). Translation involves reading the genetic code in three-letter batches, so when one or two bases are added or deleted, the reading frame is shifted. This *frameshift mutation* results in a completely different interpretation of what the code says and produces an entirely different amino acid strand. As you can imagine, these effects have disastrous consequences, because the expected gene product isn't produced. If three bases are added or deleted, the reading frame isn't affected. The result of a three-base insertion or deletion, called an *in-frame mutation,* is that one amino acid is either added (insertion) or lost (deletion). In-frame mutations can be just as bad as frameshift mutations. I cover the consequences of these sorts of mutations in the section "Facing the Consequences of Mutation" later in this chapter.

What Causes Mutation?

Mutations can occur for a whole suite of reasons. In general, though, the causes of mutations are either random or because of exposure to outside agents such as chemicals or radiation. In the sections that follow, I delve into each of these causes.

Spontaneous mutations

Spontaneous mutation occurs randomly and without any urging from some external cause. It's a natural, normal occurrence. Because the vast majority of your DNA doesn't code for anything, most spontaneous mutation goes unnoticed (check out Chapter 8 for more details about your noncoding "junk" DNA). But when mutation occurs within a gene, the function of the gene can be changed or disrupted. Those changes can then result in unwanted side effects (such as cancer, which I address in Chapter 14).

Scientists are all about counting, sorting, and quantifying, and it's no different with mutations. Spontaneous mutations are measured in the following ways:

✔ **Frequency:** Mutations are sometimes measured by the frequency of occurrence. *Frequency* is the number of times some event occurs within a group of individuals. When you hear that one in some number of persons has a particular disease-causing allele, the number is a frequency. For example, one study estimates that the X-linked disease hemophilia has a frequency of 13 cases for every 100,000 males.

✔ **Rate:** Another way of looking at mutations is in the framework of a *rate,* like the number of mutations occurring per round of cell division, or the number of mutations per gamete or per generation. Mutation rates appear to vary a lot from organism to organism. Even within a species, mutation rates vary depending on which part of the genome you're examining. Some convincing studies show that mutation rate even varies by sex and that mutation rates are higher in males than females (check out the sidebar "Dad's age matters, too" for more on this topic). Regardless of how it's viewed, spontaneous mutation occurs at a steady but very low rate (like around one per million gametes).

Most spontaneous mutations occur because of mistakes made during replication (all the details of how DNA replicates itself are in Chapter 7). Here are the three main sources of error that can happen during replication:

✔ Mismatched bases are overlooked during proofreading.

✔ Strand slip-ups lead to deletions or insertions.

✔ Spontaneous but natural chemical changes cause bases to be misread during replication, resulting in substitutions or deletions.

Mismatches during replication

Usually, *DNA polymerase* catches and fixes mistakes made during replication. DNA polymerase has the job of reading the template, adding the appropriate complementary base to the new strand, and then proofreading the new base before moving to the next base on the template. DNA polymerase can snip out erroneous bases and replace them, but occasionally, a wrong base escapes detection. Such an error is possible because noncomplementary bases can form hydrogen bonds through what's called *wobble pairing*. As you can see in Figure 13-1, wobble pairing can occur

✔ **Between thymine and guanine** without any modifications to either base (because these noncomplementary bases can sometimes form bonds in odd spots).

✔ **Between cytosine and adenine** only when adenine acquires an additional hydrogen atom (called *protonation*).

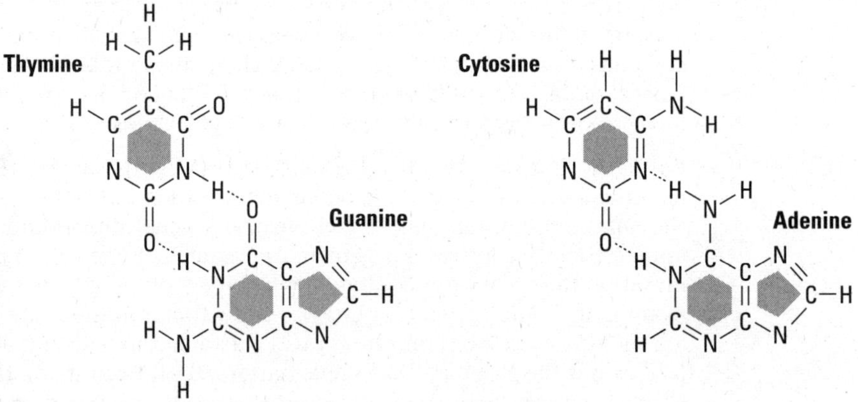

Figure 13-1:
Wobble base pairing allows mismatched bases to form bonds.

If DNA repair crews don't catch the error and fix it (see the section "Evaluating Options for DNA Repair" later in this chapter), and the mismatched base remains in place, the mistake is perpetuated after the next round of replication, apparent in Figure 13-2. The mistaken base is read as part of the template strand, and its complement is added to the newly replicated strand opposite. Thus, the mutation is permanently added to the structure of the DNA in question.

Dad's age matters, too

The relationship between maternal age and an increased incidence of chromosome problems, particularly Down syndrome, is very well-known. *Nondisjunction events* — the failure of chromosomes to separate normally during meiosis — in developing eggs are thought to be a consequence of aging in women. Very few similar genetic problems appear to arise in men, who, unlike women, produce new *gametes* (reproductive cells) in the form of sperm throughout their lifetime. However, older men are susceptible to germ-cell mutations that can cause heritable disorders in their children.

The reason that older men are more susceptible to spontaneous germ-cell mutations is

the same reason that they're less likely to have nondisjunctions — males produce sperm throughout their life. This continued sperm production means that a 50-year-old man's germ cells have replicated over 800 times. As DNA ages, replication gets less accurate, and repair mechanisms become faulty. Thus, older fathers have an increased risk (although it's still only slight) of fathering children with genetic disorders. *Achondroplasia* (an autosomal dominant form of dwarfism that's typified by shortened limbs and an enlarged head), *Marfan syndrome* (a disorder of skeletal and muscle tissue that causes heart and eye problems) and *progeria* (a disease that causes rapid aging in children) are all associated with older fathers.

Figure 13-2:
A mismatched base pair creates a permanent change in the DNA with one round of replication.

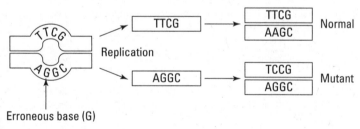

Strand slip-ups

During replication, both strands of DNA are copied more or less at the same time. Occasionally, a portion of one strand (either the template or the newly synthesized strand) can form a loop in a process called *strand slippage*. In Figure 13-3, you can see that strand slippage in the new strand results in an insertion, and slippage in the template strand results in a deletion.

Strand slippage is associated with repeating bases. When one base is repeated more than five times in a row (AAAAAA, for example), or when any number of bases are repeated over and over (such as AGTAGTAGT), strand slippage during replication is far more likely to occur. In some cases, the mistakes produce lots of variation in noncoding DNA, and the variation is useful for determining individual identity; this is the basis for DNA fingerprinting (see Chapter 18 for that discussion). When repeat sequences occur within genes, the addition of new repeats can lead to a stronger effect of the gene. This strengthening effect, called *anticipation,* occurs in genetic disorders such as Huntington disease. (You can find out more about anticipation in Chapter 4.)

Another problem that repeated bases generate is unequal crossing-over. During meiosis, homologous chromosomes are supposed to align exactly so that exchanges of information are equal and don't disrupt genes (turn to Chapter 2 for a meiosis review). Unequal crossing-over occurs when the exchange between chromosomes results in the swapping of uneven amounts of material. Repeated sequences cause unequal crossovers because so many similar bases match. The identical bases can align in multiple, matching ways that result in mismatches elsewhere along the chromosome. Unequal crossover events lead to large-scale chromosome changes (like those I describe in Chapter 15). Chromosomes in cells affected by cancer are also vulnerable to crossing-over errors (see Chapter 14 for details).

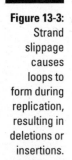

Figure 13-3: Strand slippage causes loops to form during replication, resulting in deletions or insertions.

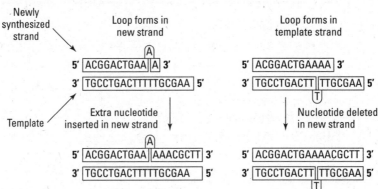

Spontaneous chemical changes

DNA can undergo spontaneous changes in its chemistry that result in both deletions and substitutions. DNA naturally loses purine bases at times in a process called *apurination.* Most often, a purine is lost when the bond between adenine and the sugar, deoxyribose, is broken. (See Chapter 6 for a reminder of what a nucleotide looks like.) When a purine is lost, replication treats the spot occupied by the orphaned sugar as if it never contained a base at all, resulting in a deletion.

Deamination is another chemical change that occurs naturally in DNA. It's what happens when an amino group (composed of a nitrogen atom and two hydrogens — NH_2) is lost from a base. Figure 13-4 shows the before and after stages of deamination. When cytosine loses its amino group, it's converted to uracil. Uracil normally isn't found in DNA at all because it's a component of RNA. If uracil appears in a DNA strand, replication replaces the uracil with a thymine, creating a substitution error. Until it's snipped out and replaced during repair (see "Evaluating Options for DNA Repair" later in this chapter), uracil acts as a template during replication and pairs with adenine. Ultimately, what was a C-G pair transitions into an A-T pair instead.

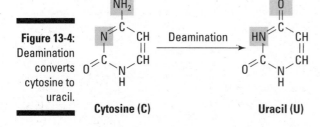

Figure 13-4: Deamination converts cytosine to uracil.

Cytosine (C)   Uracil (U)

Induced mutations

REMEMBER

Induced mutations result from exposure to some outside agent such as chemicals or radiation. It probably comes as no surprise to you to find out that many chemicals can cause DNA to mutate. *Carcinogens* (chemicals that cause cancers) aren't uncommon; the chemicals in cigarette smoke are probably the biggest offenders. In addition to chemicals that cause mutations, sources of radiation, from X-rays to sunlight, are also mutagenic. A *mutagen* is any factor that causes an increase in mutation rate. Mutagens may or may not have phenotypic effects — it depends on what part of the DNA is affected. The following sections cover two major categories of mutagens: chemicals and radiation. Each causes different damage to DNA.

Chemical mutagens

The ability of chemicals to cause permanent changes in the DNA of organisms was discovered by Charlotte Auerbach in the 1940s (see the sidebar "The chemistry of mutation" for the full story). There are many types of mutagenic chemicals; the following sections address four of the most common.

Base analogs

Base analogs are chemicals that are structurally very similar to the bases normally found in DNA. Base analogs can get incorporated into DNA during replication because of their structural similarity to normal bases. One base analog, 5-bromouracil, is almost identical to the base thymine. Most often, 5-bromouracil (also known as 5BU), which is pictured in Figure 13-5, gets incorporated as a substitute for thymine and, as such, is paired with adenine. The problem arises when DNA replicates again with 5-bromouracil as part of the template strand; 5BU is mistaken for a cytosine and gets mispaired with guanine. The series of events looks like this: 5-bromouracil is incorporated

where thymine used to be, so T-A becomes 5BU-A. After one round of replication, the pair is 5BU-G, because 5BU is prone to chemical changes that make
it a mimic of cytosine, the base normally paired with guanine. After a second
round of replication, the pair ends up as C-G, because 5BU isn't found in
normal DNA. Thus, an A-T ends up as a C-G pair.

Another class of base analog chemicals that foul up normal base pairing is
deaminators. Deamination is a normal process that causes spontaneous mutation; however, problems arise because deamination can get speeded up when
cells are exposed to chemicals that selectively knock out amino groups converting cytosines to uracils.

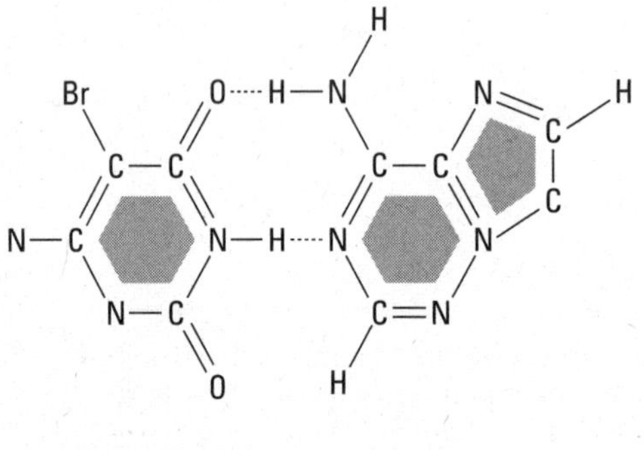

Figure 13-5:
Base
analogs,
such as 5-
bromouracil,
are very
similar to
normal
bases.

5–bromouracil   Adenine

Alkylating agents

Like base analogs, *alkylating agents* induce mispairings between bases.
Alkylating agents, such as the chemical weapon mustard gas, add chemical groups to the existing bases that make up DNA. As a consequence, the
altered bases pair with the wrong complement, thus introducing the mutation.
Surprisingly, alkylating agents are often used to fight cancer as part of chemotherapy; therapeutic versions of alkylating agents may inhibit cancer growth
by interfering with the replication of DNA in rapidly dividing cancer cells.

Free radicals

Some forms of oxygen, called *free radicals,* are unusually reactive, meaning
they react readily with other chemicals. These oxygens can damage DNA
directly (by causing strand breaks) or can convert bases into new unwanted
chemicals that, like most other chemical mutagens, then cause mispairing
during replication. Free radicals of oxygen occur normally in your body as a
product of metabolism, but most of the time, they don't cause any problems.
Certain activities — such as cigarette smoking and high exposure to radiation, pollution, and weed killers — increase the number of free radicals in
your system to dangerous levels.

Intercalating agents

Many different kinds of chemicals wedge themselves between the stacks of bases that form the double helix itself, disrupting the shape of the double helix. Chemicals with flat ring structures, such as dyes, are prone to fitting themselves between bases in a process called *intercalation*. Figure 13-6 shows intercalating agents at work. Intercalating agents create bulges in the double helix that often result in insertions or deletions during replication, which in turn cause frameshift mutations.

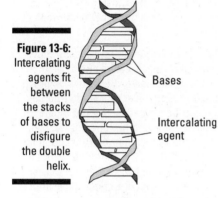

Figure 13-6: Intercalating agents fit between the stacks of bases to disfigure the double helix.

Bases

Intercalating agent

Radiation

Radiation damages DNA in a couple of different ways. First, radiation can break the strands of the double helix by knocking out bonds between sugars and phosphates (see Chapter 6 for a review of how the strands are put together). If only one strand is broken, the damage is easily repaired. But when two strands are broken, large parts of the chromosome can be lost; these kinds of losses can affect cancer cells (see Chapter 14) and cause birth defects (see Chapter 15).

Second, radiation causes mutation through the formation of *dimers*. Dimers (*di-* meaning "two"; *mer* meaning "thing") are unwanted bonds between two bases stacked on top of each other (on the same side of the helix, rather than on opposite sides). They're most often formed when two thymines in a DNA sequence bind together, which you can see in Figure 13-7.

Thymine dimers can be repaired, but if damage is extensive, the cell dies (see Chapter 14 for how cells are programmed to die). When dimers aren't repaired, the machinery of DNA replication assumes that two thymines are present and puts in two adenines. Unfortunately, cytosine and thymine can also form dimers, so the default repair strategy sometimes introduces a mutation instead.

The chemistry of mutation

If ever anyone had an excuse to give up, it was Charlotte Auerbach. Born in Germany in 1899, Auerbach was part of a lively and highly educated Jewish family. In spite of her deep interest in biology, she became a teacher, convinced that higher education would be closed to her because of her religious heritage. As anti-Jewish sentiment in Germany grew, Auerbach lost her teaching job in 1933 when every Jewish secondary-school teacher in the country was fired. As a result, she emigrated to Britain, where she earned her PhD in genetics in 1935. Charlotte Auerbach didn't enjoy the respect her degree and abilities deserved. She was treated as a lab technician and instructed to clean the cages of experimental animals. All that changed when she met Herman Muller in 1938. Like Auerbach, Muller was interested in how genes work; his approach to the problem was to induce mutations using radiation and then examine the effects produced by the defective genes. Inspired by Muller, Auerbach began work on chemical mutagens. She focused her efforts on mustard gas, a horrifically effective chemical weapon used extensively during World War I. Her research involved heating liquid mustard gas and exposing fruit flies to the fumes. It's a wonder her experiments didn't kill her.

What Charlotte's experiments did do was show that mustard gas is an alkylating agent, a mutagen that causes substitution mutations. Shortly after the end of World War II, and after persevering through burns caused by hot mustard gas, Auerbach published her findings. At last, she received the recognition and respect her work warranted. Charlotte Auerbach went on to have a long and highly successful career in genetics. She stopped working only after old age robbed her of her sight. She died in Edinburgh, Scotland, in 1994 at the age of 95.

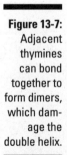
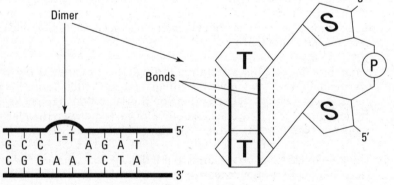

Figure 13-7:
Adjacent thymines can bond together to form dimers, which damage the double helix.

Facing the Consequences of Mutation

When a gene mutates and that mutation is passed along to the next generation, the new, mutated version of the gene is considered a new allele. *Alleles* are simply alternative forms of genes. For most genes, many alleles exist. The effects of mutations that create new alleles are compared with the mutations' physical *(phenotypic)* effects. If a mutation has no effect, it's considered *silent.* Most silent mutations result from the redundancy of the genetic code. The code is redundant in the sense that multiple combinations of bases have identical meanings (see Chapter 9 for more about the redundancy of genetic code).

Sometimes, mutations cause a completely different amino acid to be put in during translation. Mutations that actually alter the code are called *missense mutations*. A *nonsense mutation* occurs when a message to stop translation (called a *stop codon*) is introduced into the middle of the sequence. The introduction of the stop codon usually means the gene stops functioning altogether.

Mutations are often divided into two types:

✔ **Neutral:** When the amino acid produced from the mutated gene still creates a fully functional, normal protein (via translation; see Chapter 9).

✔ **Functional change:** When a new protein is created representing a change in function of the gene. A *gain-of-function mutation* creates an entirely new trait or phenotype. Sometimes, the new trait is harmless, like a new eye color. In other cases, the gain is decidedly harmful and usually autosomal dominant (flip to Chapter 12 for more on autosomal dominant traits) because the gene is producing a new protein that actually does something (the gain-of-function part). Even though there's only one copy of the new allele, its effect is noticeable and thus considered dominant over the original, unmutated allele.

If a mutation causes the gene to stop functioning altogether or vastly alters normal function, it's considered a *loss-of-function mutation*. All nonsense mutations are loss-of-function mutations, but not all loss-of-function mutations are the result of nonsense mutations. The usefulness of the protein made from a particular gene can be lost, even when no stop codon has been added prematurely. Insertions and deletions are often loss-of-function mutations because they cause frameshifts (Chapter 9 explains how the genetic code is read in frames). Frameshifts cause an entirely new set of amino acids to be put together from the new set of instructions. Most of the time, these new proteins are useless and nonfunctional. Loss-of-function mutations are usually recessive because the normal, unmutated allele is still producing product — usually enough to compensate for the mutated allele. Loss-of-function mutations are only detected when a person is homozygous for the mutation and is making no functional gene product at all.

Evaluating Options for DNA Repair

Mutations in your DNA can be repaired in four major ways:

- **Mismatch repair:** Incorrect bases are found, removed, and replaced with the correct, complementary base. Most of the time, DNA polymerase, the enzyme that helps make new DNA, immediately detects mismatched bases put in by mistake during replication. DNA polymerase can back up and correct the error without missing a beat. But if a mismatched base gets put in some other way (through strand slip-ups, for example), a set of enzymes that are constantly scrutinizing the double strand to detect bulges or constrictions signals a mismatched base pair. The mismatch repair enzymes can detect any differences between the template and the newly synthesized strand, so they clip out the wrong base and, using the template strand as a guide, insert the correct base.

- **Direct repair:** Bases that are modified in some way (like when oxidation converts a base to some new form) are converted back to their original states. Direct repair enzymes look for bases that have been converted to some new chemical, usually by the addition of some unwanted group of atoms. Instead of using a cut-and-paste mechanism, the enzymes clip off the atoms that don't belong, converting the base back to its original form.

- **Base-excision repair:** Base-excisions and nucleotide-excisions (check out the next bullet) work in much the same way. *Base-excisions* occur when an unwanted base (such as uracil; see the section "Spontaneous chemical changes" earlier in this chapter) is found. Specialized enzymes recognize the damage, and the base is snipped out and replaced with the correct one.

- **Nucleotide-excision repair:** *Nucleotide-excision* means that the entire nucleotide (and sometimes several surrounding nucleotides as well) gets removed all at once. When intercalating agents or dimers distort the double helix, nucleotide-excision repair mechanisms step in to snip part of the strand, remove the damage, and synthesize fresh DNA to replace the damaged section.

As with base excision, specialized enzymes recognize the damaged section of the DNA. The damaged section is removed, and newly synthesized DNA is laid down to replace it. In nucleotide-excision, the double helix is opened up, much like it is during replication (which I cover in Chapter 7). The sugar-phosphate backbone of the damaged strand is broken in two places to allow removal of that entire portion of the strand. DNA polymerase synthesizes a new section, and DNA ligase seals the breaks in the strand to complete the repair process.

Examining Common Inherited Diseases

Even though mutation is a common occurrence, most inherited diseases are comfortingly rare. Inherited disorders are often recessive and show up only when an individual is homozygous for the trait. Inherited diseases aren't non-existent, though. The following sections provide details on three relatively common inherited diseases. You can find out more about inheritance patterns in Chapter 12.

Cystic fibrosis

The most common inherited disorder among Caucasians in the United States is cystic fibrosis (CF). This autosomal recessive disorder occurs in roughly one in every 3,000 births (*autosomal recessive* means the gene isn't on a sex chromosome and a person must have two copies of the allele to get the disease; see Chapter 3). The mutations (there can be many) that cause CF occur in a gene located on chromosome 7. Persons affected with CF produce thick, sticky mucus in their lungs, intestines, and pancreas.

The gene implicated in CF, called the *cystic fibrosis transmembrane conductance regulator gene* (or *CFTR* for short), normally controls the passage of salt across cell membranes. Water naturally moves to areas where salt is more concentrated, so the movement of salt from one place to another has an effect on how much water is present in parts of the body. In persons with CF, the removal of salt from the body (via sweat) is abnormally high. As a result, the lungs, pancreas, and digestive system can't retain enough water to dilute the mucus normally found in those systems, so the buildup of thick mucus blocks breathing passages and makes waste elimination difficult, causing severe breathing and digestive difficulties and a high susceptibility to respiratory illnesses.

CF is diagnosed in two ways:

- Persons who may be carriers for the mutated allele can undergo genetic testing.
- Children possibly affected by the disease are diagnosed by a "sweat test." Their sweat is tested for salt content, and abnormally high amounts of salt indicate that the child has the disease.

CF is a target of gene therapy (see Chapter 16), but it resists a cure. Most afflicted persons must endure a lifetime of treatment that includes having someone pound on their chests so that they can remove the mucus from their lungs by coughing. The prognosis for CF has improved dramatically, yet most persons affected by the disease don't live far beyond their 30s.

For additional information on cystic fibrosis and to find contacts in your area, contact the Cystic Fibrosis Foundation at 800-344-4823 (`www.cff.org`) or the Canadian Cystic Fibrosis Foundation at 800-378-2233 (`www.cystic fibrosis.ca`).

Sickle cell anemia

Sickle cell anemia is the most common genetic disorder among African Americans in the United States — roughly one in every 400 births is affected by this autosomal recessive disorder. The mutation responsible for sickle cell is found on chromosome 11, the gene responsible for making one part of the protein complex that composes hemoglobin (check out Chapter 9 for how complex proteins are formed). In the case of sickle cell, one base is mutated from adenine to thymine (a *transversion*). The mistake changes one amino acid added during translation from glutamic acid to valine, producing a protein that folds improperly and can't carry oxygen effectively.

The red blood cells of persons affected by sickle cell take on the disease's characteristic crescent shape when oxygen levels in the body are lower than usual (often as the result of aerobic exercise). The sickling event has the side effect of causing blood clots to form in the smaller blood vessels (capillaries) throughout the body. Clot formation is extremely painful and also causes damage to tissues that are sensitive to oxygen deprivation. Persons with sickle cell are vulnerable to kidney failure, yet with good medical care, most affected persons live into middle adulthood (40 to 50 years of age).

For more information on sickle cell anemia, contact the American Sickle Cell Anemia Association at 216-229-8600 (`www.ascaa.org`).

Tay-Sachs disease

An autosomal recessive disorder, Tay-Sachs disease is a progressive, fatal disease of the nervous system and is unusually common among persons of *Ashkenazi* (Eastern European) Jewish ancestry. One in every 30 to 40 persons of Jewish ancestry is a carrier of Tay-Sachs disease. French Canadians and persons of Cajun (south Louisiana) descent are also often carriers of the mutated allele.

The mutation that causes Tay-Sachs disease is found in the gene that codes for the enzyme hexosaminidase A *(HEXA)*. Normally, your body breaks down a class of fats called *gangliosides*. When *HEXA* is mutated, the normal metabolism of gangliosides stops and the fats build up in the brain, causing damage. Children inheriting two copies of the affected allele are normal at birth, but as the fats build up in their brain over time, these children become blind, deaf, mentally impaired, and ultimately paralyzed. Most children with Tay-Sachs disease don't survive beyond the age of 4. Unlike some metabolic disorders, such as phenylketonuria (see Chapter 12), changes in diet don't prevent the buildup of the unwanted chemical in the body.

For more information on Tay-Sachs disease, contact the National Tay-Sachs & Allied Diseases Association at 800-906-8723 (www.ntsad.org).

Chapter 14

Taking a Closer Look at the Genetics of Cancer

*I*f you've had personal experience with cancer, you're not alone. I've lost family members, co-workers, students, and friends to this insidious disease — it's highly likely that you have, too. Second only to heart disease, cancer causes the deaths of around 560,000 persons a year in the United States alone, and it was estimated that nearly 1.5 million Americans would be diagnosed with cancer in 2009. Cancer is a genetic disorder that involves how cells grow and divide. Your likelihood of getting cancer is influenced by your genes (the genes you inherited from your parents) and your exposure to certain chemicals and radiation. Sometimes, cancers occur from random, spontaneous mutations — events that defy explanation and have no apparent cause. In this chapter, you find out what cancer is, the genetic basis of cancers, and some details about the most common types of cancer.

If you skipped over Chapter 2 on cells, you may want to backtrack before delving into this chapter, because cell information helps you understand what you read here. All cancers arise from mutations; you can discover how and why mutations occur in Chapter 13. I cover cancer treatments in the form of gene therapy in Chapter 16.

Defining Cancer

Cancer is, in essence, cell division running out of control. As I explain in Chapter 2, the cell cycle is normally a carefully regulated process. Cells grow and divide on a schedule that's determined by the type of cell involved. Skin cells grow and divide continuously because replacing dead skin cells is a never-ending job. Some cells retire from the cell cycle: The cells in your brain and nervous system don't take part in the cell cycle; no growth and no cell division occur there during adulthood. Cancer cells, on the other hand, don't obey the rules and have their own, often frightening, agendas and schedules. Table 14-1 lists the probability of developing one of the six most common cancers in the U.S.

Table 14-1	Lifetime Probability of Developing Cancer
Type of Cancer	*Risk*
Prostate	1 in 6 (16.7%)
Breast	1 in 8 (12.5%)
Colon and rectum	1 in 20 (5%)
Skin	Men: 1 in 39 (2.6%) Women: 1 in 58 (1.7%)
Oral	1 in 72 (1.4%)

In the following sections, I outline the two basic categories of tumors — benign and malignant. *Benign tumors* grow out of control but don't invade surrounding tissues. *Malignant tumors* are invasive and have a disturbing tendency to travel and show up in new sites around the body.

Benign growths: Nearly harmless growths

In a benign tumor, the cells divide at an abnormally high rate but remain in the same location. Benign tumors tend to grow rather slowly, and they create trouble because of tumor formation. In general, a *tumor* is any mass of abnormal cells. Tumors cause problems because they take up space and can compress nearby organs. For example, a tumor that grows near a blood vessel can eventually cut off blood flow just by virtue of its bulk. Benign growths can sometimes also interfere with normal body function and even affect genes by altering hormone production (see Chapter 10 for how hormones control genes).

Generally, benign growths are characterized by their lack of invasiveness. A benign tumor is usually well-defined from surrounding tissue, pushes other tissues aside, and can be easily moved about. The cells of benign tumors usually bear a strong resemblance to the tissues they start from. For example, under a microscope, a cell from a benign skin tumor looks similar to a normal skin cell.

A different sort of benign cell growth is called a *dysplasia,* a cell with an abnormal appearance. Dysplasias aren't cancerous (that is, they don't divide out of control) but are worrisome because they have the potential to go through changes that lead to malignant cancers. When examined under the microscope, dysplasias often have enlarged cell nuclei and a "disorderly" appearance. In other words, they have irregular shapes and sizes relative to other cells of the same type. Tumor cells sometimes start as one cell type (benign) but, if left untreated, can give rise to more invasive types as time goes on.

Treatment of benign growths (including dysplasias) varies widely depending on the size of the tumor, its potential for growth, the location of the growth, and the probability that cell change may lead to malignancy (invasive forms of cancer; see the following section). Some benign growths shrink and disappear on their own, and others require surgical removal.

The best defense against benign tumors (and any sort of cancer) is early detection.

- ✔ Men should undergo yearly prostate checkups beginning at age 50.

- ✔ All women over age 20 should do breast self-exams every month.

- ✔ All women should get yearly mammograms starting at age 40.

- ✔ Women should have a *Pap smear,* a test to assess the cells of their cervix, every one to three years depending on their age and the results of their previous exams.

Malignancies: Seriously scary results

Probably one of the most frightening words a doctor can utter is "malignant." *Malignancy* is characterized by cancer cells' rapid growth, invasion into neighboring tissues, and the tendency to metastasize. *Metastasis* occurs when cancer cells begin to grow in other parts of the body besides the original tumor site; cancers tend to metastasize to the bones, liver, lungs, and brain. Like benign growths, malignancies form tumors, but malignant tumors are poorly defined from the surrounding tissue — in other words, it's difficult to tell where the tumor ends and normal tissue begins. (See the section "Metastasis: Cancer on the move" later in this chapter for more info on the process.)

Malignant cells tend to look very different from the cells they arise from (Figure 14-1 shows the differences). The cells of malignant tumors often look more like tissues from embryos or stem cells than normal "mature" cells. Malignant cells tend to have large nuclei, and the cells themselves are usually larger than normal. The more abnormal the cells appear, the more likely it is that the tumor may be invasive and able to metastasize.

Normal cells

Cancer cell

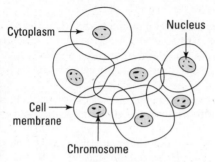
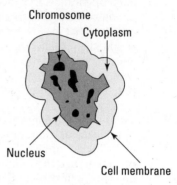

Cytoplasm

Nucleus

Cell membrane

Chromosome

Chromosome

Cytoplasm

Nucleus

Cell membrane

Figure 14-1:
Normal and
malignant
cells look
very
different.

Malignancies fall into one of five categories based on the tissue type they arise from:

- ✔ **Carcinomas** are associated with skin, nervous system, gut, and respiratory tract tissue.

- ✔ **Sarcomas** are associated with connective tissue (such as muscle) and bone.

- ✔ **Leukemias** (related to sarcomas) are cancers of the blood.

- ✔ **Lymphomas** develop in glands that fight infection (lymph nodes and glands scattered throughout the body).

- ✔ **Myelomas** start in the bone marrow.

Cancer can occur in essentially any cell of the body. The human body has 300 or so different cell types, and doctors have identified 200 forms of cancer.

Treatment of malignancy varies depending on the location of the tumor, the degree of invasion, the potential for metastasis, and a host of other factors. Treatment may include surgical removal of the tumor, surrounding tissues, and *lymph nodes* (little knots of immune tissue found in scattered locations around the body). *Chemotherapy* (administering of anticancer drugs) and radiation may also be used to combat the growth of invasive cancers. Some forms of gene therapy, which I address in Chapter 16, may also prove helpful.

Metastasis: Cancer on the move

Cells in your body stay in their normal places because of physical barriers to cell growth. One such barrier is called the *basal lamina.* The basal lamina (or basement membrane) is a thin sheet of proteins that's sandwiched between layers of cells. Metastatic cells produce enzymes that destroy the basal lamina and other barriers between cell types. Essentially, metastatic cells eat their way out by literally digesting the membranes designed to keep cells from invading each other's space. Sometimes, these invasions allow metastatic cells to enter the bloodstream, which transports the cells to new sites where they can set up shop to begin a new cycle of growth and invasion.

Another consequence of breaking down the basal lamina is that the action allows tumors to set up their own blood supply in a process called *angiogenesis*. Angiogenesis is the formation of new blood vessels to supply the tumor cells with oxygen and nutrients. Tumors may even secrete their own growth factors to encourage the process of angiogenesis. Oddly, primary tumors (the first site of tumor growth in the body) seem to restrain angiogenesis in metastasized tissue. When the primary tumor is removed, this control is released, and angiogenesis in the metastasized tumors speeds up. Increased angiogenesis means that the metastasized tumors may start to grow more rapidly, launching a new round of treatment.

A study of breast cancer in 2003 showed that cells that plant the seeds for metastasis depart the original tumor sites without the mutations that are thought to create metastasis in the first place. The wandering cells acquire mutations later, after they've settled in new locations. This discovery means that the old view (still found in many textbooks) of how metastasis develops — a stepwise, one mutation at a time process that happens in the primary tumor cells — is probably wrong.

Recognizing Cancer as a DNA Disease

Normally, a host of genes regulate the cell cycle. Thus, at its root, cancer is a disease of the DNA. Mutation damages DNA, and mutations can ultimately take the phenotype (physical trait) of cancer. The good news is it takes more than one mutation to give a cell the potential to become cancerous. The transformation from normal cell to cancer cell is thought to require certain genetic changes. These mutations can happen in any order — it's not a 1-2-3 process.

- ✔ A mutation occurs that starts cells on an abnormally high rate of cell division.

- ✔ A mutation in one (or more) rapidly dividing cells confers the ability to invade surrounding tissue.

- ✔ Additional mutations accumulate to confer more invasive properties or the ability to metastasize.

Most cancers arise from two or more mutations that occur in the DNA of *one* cell. Tumors result from many cell divisions. The original cell containing the mutations divides, and that cell's "offspring" divide over and over to form a tumor (see Figure 14-2).

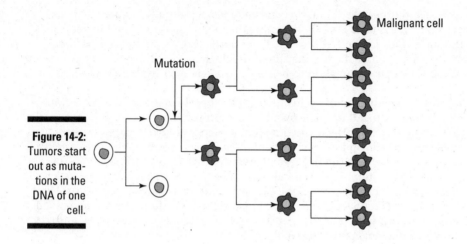

Figure 14-2:
Tumors start
out as muta-
tions in the
DNA of one
cell.

Malignant cell

Mutation

Exploring the cell cycle and cancer

The cell cycle and division (called *mitosis,* which I cover in Chapter 2) is tightly regulated in normal cells. Cells must pass through checkpoints, or stages of the cell cycle, in order to proceed to the next stage. If DNA synthesis isn't complete or damage to the DNA hasn't been repaired, the checkpoints prevent the cell from moving into another stage of division. These checkpoints protect the integrity of the cell and the DNA inside it. Figure 14-3 shows the cell cycle and the checkpoints that occur throughout.

Chapter 2 explains two checkpoints of the cell cycle. Four major conditions — basically, quality control points — must be met for cells to divide:

- ✔ DNA must be undamaged (no mismatches, bulges, or strand breaks like those described in Chapter 13) for the cell to pass from G1 of interphase into S (DNA synthesis).
- ✔ All the chromosomes must complete replication for the cell to pass out of S.
- ✔ DNA must be undamaged to start prophase of mitosis.
- ✔ Spindles required to separate chromosomes must form properly for mitosis to be completed.

If any of these conditions aren't met, the cell is "arrested" and not allowed to continue to the next phase of division. Many genes and the proteins they produce are responsible for making sure that cells meet all the necessary conditions for cell division.

When it comes to cancer and how things go wrong with the cell cycle, two types of genes are especially important:

✔ **Proto-oncogenes,** which stimulate the cell to grow and divide, basically acting to push the cell through the checkpoints

✔ **Tumor-suppressor genes,** which act to stop cell growth and tell cells when their normal life spans have ended

Figure 14-3:
Quality control points in the cell cycle protect your cells from mutations that can cause cancer.

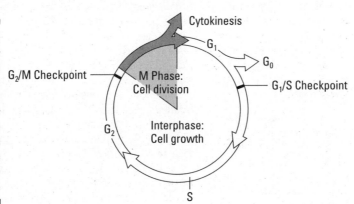

Basically, two things go on in the cell: One set of genes (and their products) acts like an accelerator to tell cells to grow and divide, and a second set of genes puts on the brakes, telling cells when to stop growing, when not to divide, and even when to die.

The mutations that cause cancer turn proto-oncogenes into either *oncogenes* (turning the accelerator permanently "on") or damage tumor-suppressor genes (removing the brakes).

Genes gone wrong: Oncogenes

You can think of oncogenes as "on" genes because that's essentially what these genes do: They keep cell division turned on. Many genes, when mutated, can become oncogenes. All oncogenes have several things in common:

✔ Their mutations usually represent a gain of function (see Chapter 13).

✔ They're dominant in their actions.

✔ Their effects cause excessive numbers of cells to be produced.

Oncogenes were the first genes identified to play a role in cancer. In 1910, Peyton Rous identified a virus that caused cancer in chickens. It took 60 years for scientists to identify the gene carried by the virus, the first known oncogene. It turns out that many viruses can cause cancer in animals and humans; for more on how these viruses do their dirty work, see the sidebar "Exploring the link between viruses and cancer."

Exploring the link between viruses and cancer

It's becoming clearer that viruses play a significant role in the appearance of cancer in humans. Second only to the risk factor of cigarette smoking, viruses are responsible for at least 15 percent of all malignancies. It turns out that viruses may alter how epigenetics controls when genes are turned on and off (you can find out more about epigenetics in Chapter 4), allowing cells to grow out of control.

One class of viruses implicated in cancer is *retroviruses*. One familiar retrovirus that makes significant assaults on human health is HIV (Human Immunodeficiency Virus), which causes AIDS (Acquired Immunodeficiency Syndrome). And if you have a cat, you may be familiar with feline leukemia, which is also caused by a retrovirus (humans are immune to this cat virus). Most retroviruses use RNA as their genetic material. Viruses aren't really alive, so to replicate their genes, they have to hijack a living cell. Retroviruses use the host cell's machinery to synthesize DNA copies of their RNA chromosomes. The viral DNA then gets inserted into the host cell's chromosome where the virus genes can be active and wreak havoc with the cell and, in turn, the entire organism. Retroviruses that cause cancer copy their oncogenes into the host cell. The oncogenes team up with additional mutations to cause cancer.

If you've ever had a wart, then you're already acquainted with the harmless version of a virus whose relatives can cause cancer. Human papilloma virus (HPV) causes genital warts and is linked to cervical cancer in women. Infection with the HPV associated with cervical cancer usually starts with *dysplasia* (the formation of abnormal but noncancerous cells). It usually takes many years for cervical cancer to develop, which it does only rarely. Nevertheless, it was expected that roughly 11,000 women in the United States alone would be diagnosed with cervical cancer in 2009. Early screening, in the form of Pap smears, for cervical cancer has improved detection and saved the lives of countless women. In 2009, reports surfaced that prostate cancer may be linked with a retrovirus (called XMRV) that is associated with cervical cancer, suggesting the possibility that some forms of prostate cancer may be caused by a sexually transmitted virus. A 2009 study published in the *British Journal of Cancer* indicates that HPV may also be associated with breast cancer — this is somewhat good news, however, since a vaccine has already been developed to ward off the virus.

Mouse mammary tumor virus (MMTV) has long been known to cause breast cancer in mice. Recent research shows that humans may also be vulnerable to MMTV. Certain kinds of breast cancers are more common in regions (such as the Middle East and Northern Africa) where a particular species of mouse (House Mouse, *Mus domesticus*) that carries MMTV is also common. These cancers tend to be very invasive and aggressive and are often accompanied by swelling and infection-like symptoms. Researchers examined breast tissue from affected women for the presence of genes similar to those of the virus. They found that North African women often carried a MMTV-like gene; many women from this region also showed other signs of having been infected with the virus. Although the link between the virus MMTV and human breast cancer is still uncertain, this and other research suggests that viruses may play significant roles in many human cancers.

You have at least 70 naturally occurring proto-oncogenes in your DNA. Normally, these genes carry out regulatory jobs necessary for normal functioning. It's only when these genes gain mutations and become oncogenes that they switch from good genes to cancer-causers. Cancer cells tend to have multiple copies of oncogenes because the genes somehow duplicate themselves in a process called *amplification.* This duplication allows those genes to have much stronger effects than they normally would.

A group of geneticists think they've figured out how cancer genes go about copying themselves. The first step in the process is formation of a *palindrome* — a DNA sequence that reads the same way forwards and backwards. In this case, the palindrome is created when a sequence gets clipped out of the DNA, flipped around, duplicated, and then inserted into the DNA (it's called an *inverted repeat*). The DNA of tumor cells has unusual numbers of palindromes. Palindromes seem to encourage more cut-and-paste duplications in the DNA around them, leading to amplification of nearby genes, like oncogenes.

The first oncogene identified in humans resides on chromosome 11. The scientists responsible for its discovery were looking for the gene responsible for bladder cancer. They took cancerous cells and isolated their DNA; then they introduced small parts of the cancer cell's DNA into a bacteria and allowed the bacteria to infect normal cells growing in test tubes. The scientists were looking for the part of the DNA present in cancer cells that would transform the normal cells into cancerous ones. The gene they found, now called *HRAS1,* was very similar to a virus oncogene that had been found in rats. The mutation that makes *HRAS1* into an oncogene affects only three bases of the genetic code (called a codon; see Chapter 9). This tiny change causes *HRAS1* to constantly send the signal "divide" to affected cells.

Since the discovery of *HRAS1,* a whole group of oncogenes has been found; they're known collectively as the *RAS* genes. All the *RAS* genes work much the same way and, when mutated, turn the cell cycle permanently "on." In spite of their dominant activities, a single mutated oncogene usually isn't enough to cause cancer all by itself. That's because tumor-suppressor genes (see the next section) are still acting to put on the brakes and keep cell growth from getting out of control.

Oncogenes aren't usually implicated in inherited forms of cancer. Most oncogenes show up as somatic mutations that can't be passed on from parent to child.

The good guys: Tumor-suppressor genes

Tumor-suppressor genes are the cell cycle's brakes. Normally, these genes work to slow or stop cell growth and to put a halt to the cell cycle. When these genes fail, cells can divide out of control, meaning that mutations in tumor-suppressor genes are loss-of-function mutations (covered in Chapter 13). Loss-of-function mutations generally only show up as phenotype when two bad copies are present — therefore, the loss of tumor-suppression means that two events have occurred to make the cells homozygous for the mutation.

The first gene recognized as a tumor suppressor is associated with cancer of the eye, called *retinoblastoma.* Retinoblastoma often runs in families and shows up in very young children. In 1971, geneticist Alfred Knudson suggested that one mutated allele of the gene was being passed from parent to child and that a mutation event in the child was required for the cancer to occur. The gene responsible, called *RB1,* was mapped to chromosome 13 and is implicated in other forms of cancer such as breast, prostate, and bone (osteosarcoma). *RB1* turns out to be a very important gene. If both copies are mutated in embryos, the mutations are lethal, suggesting that normal *RB1* function is required for survival.

RB1 regulates the cell cycle by interacting with transcription factors (I discuss transcription factors in greater detail in Chapters 9 and 11). These particular transcription factors control the expression of genes that push the cell through the checkpoint at the end of G1, just before DNA synthesis. When the proteins that *RB1* codes for (called *pRB*) are attached to the transcription factors, the genes that turn on the cell cycle aren't allowed to function. Normally, pRB and the transcription factors go through periods of being attached and coming apart, turning the cell cycle on and off. If both copies of *RB1* are mutated, then this important brake system goes missing. As a result, affected cells move through the cell cycle faster than normal and divide without stopping. *RB1* not only interacts with transcription factors to control the cell cycle; it's also thought to play a role in replication, DNA repair, and *apoptosis* (programmed cell death).

One of the most important tumor-suppressor genes identified to date is *TP53,* found on chromosome 17, which codes for the cell-cycle regulating protein p53. Mutations that lead to loss of p53 function are implicated in a wide variety of cancers. The most important of p53's roles may be in regulating when cells die, a process known as *apoptosis:*

✔ When DNA has been damaged, the cell cycle is stopped to allow repairs to be carried out.

✔ If repair isn't possible, the cell receives the signal to die (apoptosis).

If you've ever had a bad sunburn, then you have firsthand experience with apoptosis. Apoptosis, also known by the gloomy moniker "programmed cell death," occurs when the DNA of a cell is too damaged to be repaired. Rather than allow the damage to go through replication and become cemented into the DNA as mutation, the cell voluntarily dies. In the case of your severe sunburn, the DNA of the exposed skin cells was damaged by the sun's radiation. In many cases, the DNA strands were broken, probably in many different places. Those skin cells killed themselves off, resulting in the unpleasant skin peeling that you suffered. When your DNA gets damaged from too much sun exposure or because of any other mutagen (see Chapter 13 for examples), a protein called *p21* stops the cell cycle. Encoded by a gene on the X chromosome, p21 is produced when the cell is stressed. The presence of p21 stops the cell from dividing and allows repair mechanisms to heal the damaged DNA. If the damage is beyond repair, the cell may skip p21 altogether. Instead, the tumor-suppressor protein p53 signals the cell to kill itself.

When the cell gets the message that says, "Die!" a gene called *BAX* swings into action. *BAX* sends the cell off to its destruction by signaling the mitochondria — those energy powerhouses of the cell — which release a wrecking-crew of proteins that go about breaking up the chromosomes and killing the cell from the inside. When your cells die due to injury (like a burn or infection), the process is a messy one: The cells explode, causing surrounding cells to react in the form of inflammation. Not so in apoptosis. The cells killed by the actions of apoptosis are neatly packaged so that surrounding tissues don't react. Cells that specialize in garbage collection and disposal, called *phagocytes* (meaning cells that eat), do the rest.

Drugs used to fight cancer often try to take advantage of the apoptosis pathway to cell death. The drugs turn on the signals for apoptosis to trick the cancer cells into killing themselves. Radiation therapy, also used to treat cancer by introducing double-strand breaks (see Chapter 13 for more on this kind of DNA damage), relies on the cells knowing when to die. Unfortunately, some of the mutations that create cancer in the first place make cancer cells resistant to apoptosis. In other words, in addition to growing and dividing without restriction, cancer cells don't know when to die.

Demystifying chromosome abnormalities

Large-scale chromosome changes — the kinds that are visible when karyotyping (chromosome examination; see Chapter 15 for details) is done — are associated with some cancers. These chromosome changes (like losses of chromosomes) often occur after cancer develops and occur because the DNA in cancer cells is really unstable and prone to lots of breakage. Normally, damaged DNA is detected by proteins that keep tabs on the cell cycle. When breaks are found, either the cell cycle is stopped and repairs are initiated or the cell dies. Because the root of cancer is the loss of genetic quality control functions provided by proto-oncogenes and tumor-suppressor genes, it's no surprise that breaks in the cancer-cell DNA lead to losses and rearrangements of big chunks of chromosomes as the cell cycle rolls on without interruption. One of the biggest problems with all this genetic instability in cancer cells is that a tumor is likely to have several different genotypes among its many cells, which makes treatment difficult. Chemotherapy that's effective at treating cells with one sort of mutation may not be useful for another.

Three types of damage — deletions, inversions, and translocations — can interrupt tumor-suppressor genes, rendering them nonfunctional. Translocations and inversions may change the positions of certain genes so that the gene gets regulated in a new way (see Chapter 11 for more about how gene expression is regulated by location). Chronic myeloid leukemia, for example, is caused by a translocation event between chromosomes 9 and 22. This form of leukemia is a cancer of the blood that affects the bone marrow.

Translocations generally result from double-strand breaks (radiation and cigarette smoking are risk factors). In the case of chronic myeloid leukemia, the translocation event makes chromosome 22 unusually short. (This shortened version of the chromosome is called the *Philadelphia chromosome* because geneticists working in that city discovered it.) The translocation event causes two genes, one from each chromosome, to become fused together. The new gene product acts as a powerful oncogene, leading to out-of-control cell division and eventually leukemia.

Certain cancers seem prone to losing particular chromosomes altogether, resulting in monosomies (similar to those described in Chapter 15). For instance, one copy of chromosome 10 often goes missing in the cells of glioblastomas, a deadly form of brain cancer. Cancer cells are also prone to nondisjunction leading to localized trisomies. It appears that mutations in the p53 gene, which can stop the cell cycle for DNA repair and signal apoptosis, are linked to these localized changes in chromosome number.

Breaking Down the Types of Cancers

Around 200 different cancers occur in humans. Many are site specific, meaning the tumor is associated with a particular part of the body. Some cancers seem to appear just about anywhere, in any organ system. This section isn't meant to provide an exhaustive list of cancers; instead, it touches on the genetics of some of the more common cancers.

For more information on all types of cancers, visit the American Cancer Society (www.cancer.org) and the National Cancer Institute (www.cancer.gov) online.

Hereditary cancers

Hereditary cancers are those that tend to run in families. No one ever inherits cancer; what's inherited is the predisposition to certain sorts of cancer. What this means is that certain cancers tend to run in families because one or more mutations are being passed on from parent to child. Most geneticists agree that additional mutations are required to trigger the actual disease. Just because you have a family history of a particular cancer doesn't mean you'll get it. The opposite is also true: Just because you don't have a family history doesn't mean you won't get cancer.

Prostate cancer

The most common cancer in the United States is prostate cancer. The prostate is a walnut-sized gland found at the base of a man's urinary bladder. The urethra, the tube that carries urine outside the body, runs through the center of the prostate gland. The prostate generates seminal fluid, important for the production of sperm. On average, over 200,000 men are likely to be diagnosed with prostate cancer each year. The highest rates of death from prostate cancer occur among African American men, likely because of lack of screening and delayed treatment.

Many mutations are associated with a family history of prostate cancer, but the number one risk factor associated with prostate cancer is age. Older men are far more likely to develop this disease.

For most men, the first clue of changes in the prostate gland is difficulty in urination and decreased urine flow. Many older men experience swelling of the prostate, and those changes are often benign. The best screening tests for prostate cancer are a blood test called the PSA (for prostate-specific antigen) and a manual examination by a physician. Men should start getting screened for prostate cancer at age 50. Men with a family history of the disease (father, brother, or son) should start earlier — the American Cancer Society suggests that screening begin at age 45.

Numerous genes are implicated in prostate cancer. One gene, *PRCA1* on chromosome 1, is designated "the" hereditary prostate cancer gene. But less than 10 percent of all cases of prostate cancers are thought to originate with mutations at *PRCA1*. Online Mendelian Inheritance in Man (see Chapter 24) lists at least 16 genes associated with prostate cancer, including p53 and *RB;* it's likely that several genes interact to cause the cell cycle of the prostate gland to spiral out of control. There also seems to be a link between prostate cancer and the two *BRCA* genes implicated in breast cancer. Thus, men and women with family histories of either disease may be susceptible to developing cancer. New evidence points to a viral cause as well — see the earlier sidebar "Exploring the link between viruses and cancer" in this chapter to find out more.

Breast cancer

Breast cancer is the second most common cancer in America (refer to Table 14-1). Sadly, over 40,000 people, mostly women, are likely to die of the disease each year. Different sorts of breast cancer are distinguished by the part of the breast that develops the tumor. Regardless of the type of breast cancer, though, the number one risk factor appears to be a family history of the disease. Family history of breast cancer is usually defined as having one of the following:

✔ A mother or sister diagnosed with breast or ovarian cancer before age 50

✔ Two first-degree relatives (mother, sister, daughter) on the same side of the family with breast cancer at any age

✔ A male relative diagnosed with breast cancer

Generally, the first symptom of breast cancer is a lump in the breast tissue. The lump may be painless or sore, hard (like a firm knot) or soft; the edges of the lump may not be easy to detect, but in some cases they're very easy to feel. Other symptoms include swelling, changes in the skin of the breast, nipple pain or unexpected discharge, and a swelling in the armpit.

Researchers have identified two breast cancer genes: *BRCA1* and *BRCA2* (for BReast CAncer genes 1 and 2). These genes account for slightly less than 25 percent of inherited breast cancers, however. Mutations in the gene for p53, along with numerous other genes, are also associated with hereditary forms of breast cancer (see "The good guys: Tumor-suppressor genes" for more on p53). Breast cancers associated with mutations of *BRCA1* and/or *BRCA2* seem to be inherited as *autosomal dominant disorders* (genetic disorders resulting from one bad copy of a gene; see Chapter 12 for more on inheritance patterns).

When it comes to breast cancer, penetrance is roughly 50 percent, meaning 50 percent of the people inheriting a mutation in one of the breast cancer genes will develop cancer. (This penetrance value is based on a life span of 85 years, by the way, so people living 85 years have a 50 percent chance of expressing the phenotype of cancer.)

Other cancers are also associated with mutations in *BRCA1* and *BRCA2*, including ovarian, prostate, and male breast cancer.

Both *BRCA* genes are tumor-suppressor genes. The roles these genes play in the cell cycle aren't especially well defined. *BRCA1* has a role in regulating when cells pass through the critical G1-S checkpoint, but exactly how *BRCA1* does its job isn't clear. As for *BRCA2*, it apparently has some cell cycle duties and also plays a role in DNA repair, especially of double-strand breaks.

Early detection of breast cancer is the best defense against the disease. Women with a family history of breast cancer should be screened by a physician at least once a year (some doctors recommend screenings every six months). Genetic tests are available to confirm the presence of mutations that are associated with the development of breast cancer, but at present, these tests are very expensive and don't yield complete information about the true likelihood of getting the disease. After breast cancer is diagnosed, treatment options vary based on the kind of cancer. Breast cancer is considered very treatable, and the prognosis for recovery is very good for most patients. There are hopes for a vaccine to prevent some forms breast cancer altogether; see the "Exploring the link between viruses and cancer" sidebar earlier in this chapter.

Colon cancer

One hereditary cancer that's considered highly treatable (when detected early) is colon cancer. Your colon is defined by the large intestine, the bulky tube that carries waste products to your rectum for defecation. Over 100,000 people are likely to be diagnosed with colon cancer each year. Numerous risk factors are associated with colon cancer, including:

✔ Family history of the disease (meaning parent, child, or sibling)

✔ Age; persons over 50 are at greater risk

✔ High-fat diet

✔ Obesity

✔ History of alcohol abuse

✔ Smoking

Almost all colon cancers start as benign growths called *polyps*. These polyps are tiny wart-like protrusions on the wall of the colon. If colon polyps are left untreated, a *RAS* oncogene often becomes active in the cells of one or more of the polyps, causing the affected polyps to increase in size (see the "Genes gone wrong: Oncogenes" section earlier in the chapter for more on how oncogenes work). When the tumors get big enough, they change status and are called *adenomas*. Adenomas are benign tumors but are susceptible to mutation, often of the tumor-suppressor gene that controls p53. When p53 is lost through mutation, the adenoma becomes a *carcinoma* — a malignant and invasive tumor.

Early detection and treatment is critical to prevent colon polyps from becoming cancerous. If large numbers of polyps develop, the likelihood that at least one will become malignant is very high. The good news is that the changes in the colon usually accumulate slowly, over the course of several years. The American Cancer Society recommends that all persons over 50 years of age be screened for colon cancer. Two tests are generally done: a test to detect blood in the feces and a visual inspection, called a *colonoscopy,* of the inside of the colon using a flexible scope. The test kit to detect blood in the feces is available over the counter at most drug stores. Positive results are nothing to panic over — just see your physician. A colonoscopy is carried out under light anesthesia and gives your physician the most accurate means of diagnosing the presence of polyps and grabbing samples of cells for testing.

Preventable cancers

Preventable cancers are cancers associated with particular risk factors that can be controlled and avoided. No one ever chooses to get cancer, but the lifestyle choices that people make leave them more likely to develop certain kinds of cancer in their lifetime. Three of the most avoidable kinds of cancer associated with lifestyle choices are lung cancer, mouth cancers, and skin cancer.

Lung cancer

More people die from lung cancer every year than any other kind of cancer. Over 210,000 Americans are likely to be diagnosed with lung cancer in 2010, and it's estimated that roughly 160,000 people in the United States will die from the disease in that year. Ninety percent of people who get lung cancer do so because of cigarette smoking. Let me repeat that: *90 percent* of lung cancer is associated with cigarette smoking. This statistic makes lung cancer the most preventable cancer of all.

The average age for lung cancer diagnosis is age 60. Sadly, after a patient is diagnosed with lung cancer, the prognosis is generally poor. Survival estimates vary depending on the type of lung cancer, but in general, only 20 percent of people afflicted survive longer than one year after diagnosis. That's the bad news. The good news is that if you stop smoking at any age, your lungs heal, and your risk of developing cancer goes down.

The two main types of lung cancer are both associated with tobacco use:

- ✔ **Small-cell lung cancers,** which comprise roughly 25 percent of all lung cancers, are the worst type. Named for the small, round cells that comprise these tumors, they're invasive, highly prone to metastasis, and very hard to treat.

- ✔ **Non-small-cell lung cancers** are more amenable to treatment, especially when diagnosed early.

Both types of lung cancer have similar primary symptoms: weight loss, hoarseness, a cough that won't go away, and difficulty breathing. Another symptom that's often overlooked is finger clubbing. Finger clubbing is a condition in which the tips of the fingers get wider than normal. It's a common sign of lung disease and an indication that small blood vessels aren't getting enough oxygen.

Many mutations are associated with lung cancers. Both oncogenes and tumor-suppressor genes are implicated. Almost all lung cancers involve mutations of the p53 gene — the tumor-suppressor gene that controls, among other things, programmed cell death. A *RAS* oncogene, *KRAS,* is frequently mutated in certain kinds of lung cancers. Finally, large-scale deletions of chromosomes, most often involving chromosome 3, are associated with virtually all small-cell lung cancers (see the section "Demystifying chromosome abnormalities" for more details).

Cancers of the mouth

The use of smokeless tobacco (snuff and chewing tobacco) is associated with cancers of the mouth. Roughly 7,000 persons each year die of preventable mouth cancers; men are twice as likely to get mouth cancer as women. Like lung cancer, the prognosis for persons diagnosed with mouth cancer is poor.

Only slightly more than 50 percent of persons survive beyond five years after diagnosis.

The reason the prognosis for mouth cancer is so poor is that early stages of the disease show no symptoms. Therefore, most people are unaware of the problem until the disease is more advanced. Symptoms of mouth cancer include sores on the gums, tongue, or the roof of the mouth that don't heal; lumps in the mouth; thickening of the cheek lining; and persistent mouth pain. Regular dental care helps increase early detection, improving the chance of survival.

Mutations associated with mouth cancers are often large-scale chromosome abnormalities. Cells of the mouth appear especially vulnerable to mutational losses of parts of chromosomes 3, 9, and 11 — all of which are recognized fragile sites (see Chapter 15). Oncogenes in the *RAS* family and the p53 gene are also implicated in most forms of mouth cancer.

Skin cancer

Each year, nearly 60,000 people in the United States are diagnosed with *melanoma,* a form of skin cancer. Although a predisposition to skin cancer may be inherited, the number one risk factor for skin cancer is exposure to ultraviolet light. Ultraviolet light sources include the sun and tanning booths. People with pale skin, light-colored eyes (blue or green), and fair hair are most vulnerable to ultraviolet light and thus skin cancer. If you burn easily and don't tan readily, you're at higher risk. The best way to prevent skin cancer is to stay out of the sun. If you must be exposed to the sun, *always* use sunblock with an SPF (Sun Protection Factor) higher than 30.

Sunburn is strongly associated with the development of skin cancer at a later time because radiation tends to cause double-strand DNA breaks and also glues adjacent bases in the DNA together, forming spots called dimers (see Chapter 13 for more details on these sorts of DNA damage). Damage to DNA is often so great after severe sun exposure that large numbers of skin cells die. Take a look at the section on tumor-suppressor genes earlier in this chapter to find out about the process of "programmed cell death." But some damaged DNA may escape the repair or cell death process, yielding dangerous mutations. Regular screening, the key to early detection of skin cancer, is as simple as inspecting your skin using a mirror. Look closely at all moles and freckles; asymmetrical, blotchy, or large (bigger than a pencil eraser) growths should be pointed out to your physician.

Chapter 15

Chromosome Disorders: It's All a Numbers Game

....

....

The study of chromosomes is, in part, the study of cells. Geneticists who specialize in *cytogenetics,* the genetics of the cell, often examine chromosomes as the cell divides because that's when the chromosomes are easiest to see. Cell division is one of the most important activities that cells undergo; it's required for normal life, and a special sort of cell division prepares sex cells for the job of reproduction. Chromosomes are copied and divvied up during cell division, and getting the right number of chromosomes in each cell as it divides is critical. Most chromosome disorders (such as Down syndrome) occur because of mistakes during *meiosis* (the cell division that makes sex cells; see Chapter 2).

This chapter helps you understand how and why chromosome disorders occur. You find out some of the ways geneticists study the chromosome content of cells. Knowing chromosome numbers allows scientists to decode the mysteries of inheritance, especially when the number of chromosomes (called *ploidy*) gets complicated. Counting chromosomes also allows doctors to determine the origin of physical abnormalities caused by the presence of too many or too few chromosomes.

If you skipped over Chapter 2, you may want to flip back to it before reading this chapter to get a handle on the basics of chromosomes and how cells divide.

What Chromosomes Reveal

One way a geneticist counts chromosomes is with the aid of microscopes and special dyes to see the chromosomes during *metaphase* — the one time in the cell cycle when the chromosomes take on a fat, easy-to-see, sausage shape. (Jump to Chapter 2 to review the cell cycle.) Here's how the process of examining chromosomes works:

1. A sample of cells is obtained. Almost any sort of dividing cell works as a sample, including root cells from plants, blood cells, or skin cells.

2. The cells are *cultured* — given the proper nutrients and conditions for growth — to stimulate cell division.

3. Some cells are removed from the culture and treated to stop mitosis during metaphase.

4. Dyes are added to make the chromosomes easy to see.

5. The cells are inspected under a microscope. The chromosomes are sorted, examined for obvious abnormalities, and counted.

This process of chromosome examination is called *karyotyping*. A karyotype reveals exactly how many chromosomes are present in a cell, along with some details about the chromosomes' structure. Scientists can only see these details by staining the chromosomes with special dyes.

When examining a karyotype, a geneticist looks at each individual chromosome. Every chromosome has a typical size and shape; the location of the centromere and the length of the chromosome arms (the parts on either side of the centromere) are what define each chromosome's physical appearance (refer to Chapter 2 to see what some chromosomes look like up close). The two types of chromosome arms are the

✔ *p* **arm:** The shorter of the two arms (from the word *petite,* French for "small")

✔ *q* **arm:** The longer arm (because *q* follows *p* alphabetically)

In some disorders, one of the chromosome arms is misplaced or missing. Therefore, geneticists often refer to the chromosome number along with the letter *p* or *q* to communicate which part of the chromosome is affected.

Counting Up Chromosomes

Ploidy sounds like some bizarre, extraterrestrial, science-fiction creature, but the word actually refers to the number of chromosomes a particular organism has. Two sorts of "ploidys" are commonly bandied about in genetics:

✔ **Aneuploid** refers to an imbalance in the number of chromosomes. Situations involving aneuploidy are often given the suffix *-somy* to communicate whether chromosomes are missing *(monosomy)* or extra *(trisomy).*

✔ **Euploid** refers to the number of *sets* of chromosomes an organism has. Thus, *diploid* tells you that the organism in question has two sets of chromosomes (often written as *2n,* with *n* being the haploid number of chromosomes in the set; see Chapter 2 for more on how chromosomes are counted up). When an organism is euploid, its total number of chromosomes is an exact multiple of its haploid number *(n)*.

Aneuploidy: Extra or missing chromosomes

Shortly after Thomas Hunt Morgan discovered that certain traits are linked to the X chromosome (see Chapter 5 for the full story), his student Calvin Bridges discovered that chromosomes don't always play by the rules. The laws of Mendelian inheritance depend on the segregation of chromosomes — an event that takes place during the first phase of meiosis (see Chapter 2 for meiosis coverage). But sometimes chromosomes don't segregate; two or more copies of the same chromosome are sent to one *gamete* (sperm or egg), leaving another gamete without a copy of one chromosome. Through his study of fruit flies, Bridges discovered the phenomenon of *nondisjunction,* the failure of chromosomes to segregate properly. Figure 15-1 shows nondisjunction at various stages of meiotic division. (For more on how Morgan and Bridges made their discoveries, check out the sidebar "Flies!")

While studying eye color in flies (flip to Chapter 5 for more about this X-linked trait), Bridges crossed white-eyed female flies with red-eyed males. He expected to get all white-eyed sons and all red-eyed daughters from this sort of monohybrid cross (Chapter 3 explains monohybrid crosses). But every so often, he got red-eyed sons and white-eyed daughters. Bridges already knew that females get two copies of the X chromosome and males get only one, and that eye color is linked with X. He also knew that eye color is a recessive trait; the only way females could have white eyes is to have two copies of X that both have the allele for white. So how could the odd combinations of sex and eye color that Bridges saw occur?

Bridges realized that the X chromosomes of some of his female parent flies must not be obeying the rules of segregation. During the first round of meiosis, the homologous pairs of chromosomes should separate. If that doesn't happen, some eggs get two copies of the mother's X chromosome (see Figure 15-1). In Bridges's research, both copies of the mother's X carried the allele for white eyes. When a red-eyed male fertilized a two-X egg, two results were possible, as you can see in Figure 15-2. An XXX zygote resulted in a red-eyed daughter (which usually died). An XXY zygote turned out to be a white-eyed female (check out Chapter 5 for how sex is determined in fruit flies). Fertilized eggs that had no X chromosome resulted in a red-eyed male (with genotype X). Eggs that didn't get an X chromosome and receive a Y from the father were never viable at all.

Many human chromosomal disorders arise from a sort of nondisjunction similar to that of fruit flies. For more information on these disorders, take a look at the section "Exploring Chromosome Variations" later in this chapter.

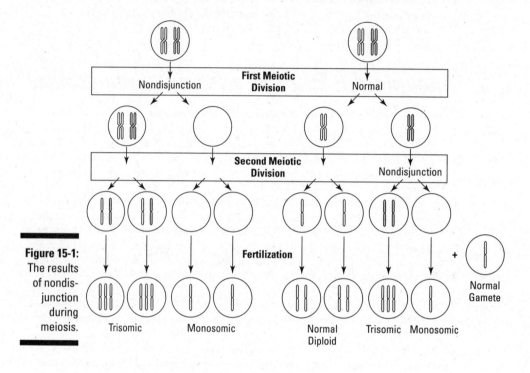

Figure 15-1:
The results of nondisjunction during meiosis.

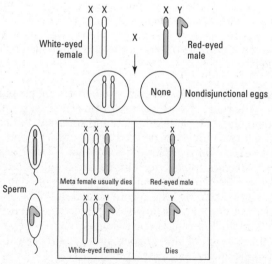

Figure 15-2:
How nondisjunction of the X chromosomes works in fruit flies.

HISTORIC ROOTS

Flies!

Some of the greatest scientific discoveries have been made in the humblest of settings. Take Thomas Hunt Morgan's laboratory, affectionately known as the Fly Room. A mere 368 square feet, it was crammed with eight students, their desks, hundreds of glass milk bottles full of fruit flies, and large bunches of bananas hung from the ceiling as food for the fruit flies. The room reeked of rotting bananas, literally buzzed with escapee flies, and had more than its fair share of cockroaches. Yet from 1910 to 1930, this cramped setting was home to some of the most important scientific discoveries of its time — discoveries that still apply to the understanding of genetics today.

Calvin Bridges and Alfred Sturtevant were both undergraduates at Columbia University in New York City in 1909. After hearing a lecture presented by Morgan, both Bridges and Sturtevant landed desk space in the Fly Room. Gregor Mendel's work had only just been rediscovered, so it was an exciting time for genetics. Fruit flies made perfect study organisms to test all the latest ideas, so the men of the Fly Room (collaborator Nettie Stevens was at the Carnegie Institution) spent hours discussing the latest publications and their own research findings. After one such discussion, Sturtevant rushed home to work up his latest idea: a map of the genes on the X chromosome. Sturtevant created his chromosome map — still accurate to this day — when he was just 20 and still an undergraduate. Bridges, at the ripe old age of 24, went on to discover nondisjunction of fly chromosomes — definitive proof that Morgan's theory of chromosomal inheritance was correct.

Euploidy: Sets of chromosomes

Every species has a typical number of chromosomes revealed by its karyotype. For example, humans have 46 total chromosomes (humans are diploid, $2n$, and $n = 23$). Your dog, if you have one, is also diploid and has 78 total chromosomes, while house cats have $2n = 38$. Chromosome number isn't very consistent, even among closely related organisms. For example, despite their similar appearance, two species of Asian deer are both diploid but have very different chromosome numbers: One species has 23 chromosomes, and the other has 6.

Many organisms have more than two sets of chromosomes (a single set of chromosomes referred to by the n is the haploid number) and are therefore considered *polyploid*. Polyploidy is rare in animals but not unheard of. Plants, on the other hand, are frequently polyploid. The reason that polyploidy is rare is sexual reproduction. Most animals reproduce sexually, meaning each individual produces eggs or sperm that unite to form zygotes that grow into offspring. An equal number of chromosomes must be allotted to each gamete for fertilization and normal life processes to occur. When an individual, such as a plant, is polyploid (particularly odd numbers like $3n$), most of its gametes wind up with an unusual number of chromosomes. This imbalance in the number of chromosomes results, functionally, in sterility (see the sidebar "Stubborn chromosomes" for more details).

Stubborn chromosomes

Horses are diploid and have 64 chromosomes. Donkeys, which are also diploid, are closely related to horses but have only 62 chromosomes. When a horse mates with a donkey, the result is a mule. These horse-donkey hybrids are larger versions of horses and have big ears and a famously stubborn disposition. Mules are usually sterile because the ploidies of horses and mules (or of donkeys and mules) are a poor match. Genetically, mules have 32 horse chromosomes and 31 donkey chromosomes, giving them a total of 63 chromosomes altogether and the odd chromosome number of $2n = 63$ — that's diploid but not euploid. When meiosis takes place, the homologous chromosomes should pair up and then segregate. During meiosis in mules, however, chromosomes often come together in groups of three, five, or six. As a result, mule gametes don't get a full complement of chromosomes and aren't viable to be fertilized. So how can any mule be a parent?

That's what the owners of a mule named Krause must have wondered in 1984 when she unexpectedly produced a foal — named Blue Moon because of the rarity of mule parenthood. Krause cohabitated with a male donkey, but genetic analysis revealed that Blue Moon had a mule genotype: 63 chromosomes that were half horse and half donkey. Apparently, when Krause's cells underwent meiosis, her horse chromosomes all segregated together. This is an outrageously improbable outcome — on the order of one in 4 billion! Even more amazingly, Krause had a second foal with the same horse-donkey genotype, meaning she produced a second egg with all horse chromosomes.

The only other way a mule can be a "parent" is via cloning, which I cover in Chapter 20. Idaho Gem, the first mule clone, was born in 2003.

Plants sometimes get around the problem of polyploidy (and its corresponding sterility) through a process called *apomixis*. Part of meiosis, apomixis results in an egg with a full complement of chromosomes. Eggs produced via apomixis can form seeds without being fertilized and therefore can produce new plants from seed. Dandelions, those hardy, persistent weeds known to all gardeners, reproduce using apomixis. Dandelions have $n = 8$ chromosomes that can come in sets of two ($2n = 16$), three ($3n = 24$), or four ($4n = 32$).

Many commercial plants are polyploid because plant breeders discovered that polyploids often are much larger than their wild counterparts. Wild strawberries, for instance, are diploid, tiny, and very tart. The large, sweet strawberries you buy in the grocery store are actually octaploid, meaning they have eight sets of chromosomes (that is, they're $8n$). Cotton is tetraploid ($4n$), and coffee can have as many as eight sets of chromosomes, while bananas are often triploid ($3n$). Many of these polyploids came about naturally and, after being discovered by plant breeders, were cultivated from cuttings (and other nonsexual plant propagation methods).

Not all polyploids are sterile. Those that result from crosses of two different species (called *hybridization*) are often fertile. The chromosomes of hybrids may have less trouble sorting themselves out during meiosis, allowing for normal gamete formation to take place. One famous animal example of a rarely fertile hybrid is a horse-donkey cross that results in a mule. Take a look at the "Stubborn chromosomes" sidebar for more information.

Exploring Chromosome Variations

Chromosomal abnormalities, in the form of aneuploidy (see the earlier section "Aneuploidy: Extra or missing chromosomes"), are very common among humans. Roughly 8 percent of all conceptions are aneuploid, and it's estimated that up to half of all miscarriages happen because of some form of chromosome disorder. Sex chromosome disorders are the most commonly observed type of aneuploidy in humans (flip to Chapter 5 for more on sex chromosomes) because X-chromosome inactivation allows individuals with more than two X chromosomes to compensate for the extra "doses" and survive the condition.

Four common categories of aneuploidy crop up in humans:

- **Nullisomy:** Occurs when a chromosome is missing altogether. Generally, embryos that are nullisomic don't survive to be born.

- **Monosomy:** Occurs when one chromosome lacks its homolog.

- **Trisomy:** Occurs when one extra copy of a chromosome is present.

- **Tetrasomy:** Occurs when four total copies of a chromosome are present. Tetrasomy is extremely rare.

Most chromosome conditions are referred to by category of aneuploidy followed by the number of the affected chromosome. For example, trisomy 13 means that three copies of chromosome 13 are present.

When chromosomes go missing

Monosomy (when one chromosome lacks its homolog) in humans is very rare. The majority of embryos with monosomies don't survive to be born. For liveborn infants, the only autosomal monosomy reported in humans is monosomy 21. Signs and symptoms of monosomy 21 are similar to those of Down syndrome (covered later in this section). Infants with monosomy 21 often have numerous birth defects and rarely survive for longer than a few days or weeks. The other monosomy commonly seen in children is monosomy of the X chromosome. Children with this condition are always female and usually lead normal lives. For more on monosomy X (also known as Turner syndrome), see Chapter 5. Monosomy 21 is the result of nondisjunction during meiosis (see the section "Aneuploidy: Extra or missing chromosomes" earlier in this chapter).

Many monosomies are partial losses of chromosomes, meaning that part (or all) of the missing chromosome is attached to another chromosome. Movements of parts of chromosomes to other, nonhomologous chromosomes are the result of *translocations*. I cover translocations in more detail in the section "Translocations" later in this chapter.

Finally, monosomies can occur in cells because of mistakes that occur during cell division (mitosis). Many of these monosomies are associated with chemical exposure and various sorts of cancers. Chapter 14 covers cell monosomies and cancer in detail.

When too many chromosomes are left in

Trisomies (when one extra copy of a chromosome is present) are the most common sorts of chromosomal abnormalities in humans. The most common trisomy is Down syndrome, or trisomy 21. Other, less common trisomies include trisomy 18 (Edward syndrome), trisomy 13 (Patau syndrome), and trisomy 8. All these trisomies are usually the result of nondisjunction during meiosis.

Down syndrome

Trisomy of chromosome 21, commonly called *Down syndrome,* affects between 1 in 600 to 1 in 800 infants. People with Down syndrome have some rather stereotyped physical characteristics, including distinct facial features, altered body shape, and short stature. Individuals with Down syndrome usually have mental retardation and often have heart defects. Nevertheless, they often lead fulfilling and active lives well into adulthood.

One of the most striking features of Down syndrome (and trisomies in general) is the precipitous increase in the number of Down syndrome babies born to mothers over 35 (see Figure 15-3). Women between 18 and 25 have a very low risk of having a baby with trisomy 21 (roughly 1 in 2,000). The risk increases slightly but steadily for women between 25 and 35 (about 1 in 900 for women 30 years old) and then jumps dramatically. By the time a woman is 40, the probability of having a child with Down syndrome is 1 in 100, and by the age of 50, the probability of conceiving a Down syndrome child is 1 in 12. Why does the risk of Down syndrome increase in the children of older women?

The majority of Down syndrome cases seem to arise from nondisjunction during meiosis. The reason behind this failure of chromosomes to segregate normally in older women is unclear. In females, meiosis actually begins in the fetus (flip to Chapter 2 for a review of gametogenesis in humans). All developing eggs go through the first round of prophase, including recombination. Meiosis in future egg cells then stops in a stage called *diplotene,* the stage of crossing-over, where homologous chromosomes are hooked together and are in the process of exchanging parts of their DNA. Meiosis doesn't start back up again until a particular developing egg is going through the process of ovulation. At that point, the egg completes the first round of meiosis and then halts again. When sperm and egg unite, the nucleus of the egg cell finishes meiosis just before the nuclei of the sperm and egg fuse to complete the process of fertilization. (In human males, meiosis begins in puberty, is ongoing, and continues without the pauses that occur in females.)

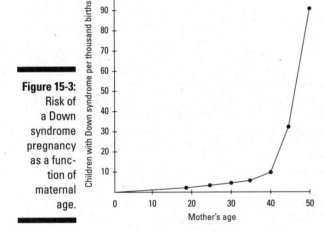

Figure 15-3:
Risk of
a Down
syndrome
pregnancy
as a func-
tion of
maternal
age.

Roughly 75 percent of the nondisjunctions responsible for Down syndrome occur during the first phase of meiosis. Oddly, most of the chromosomes that fail to segregate seem also to have failed to undergo crossing-over, suggesting that the events leading up to nondisjunction begin early in life. Scientists have proposed a number of explanations for the cause of nondisjunction and its associated lack of crossing-over, but they haven't reached an agreement about what actually happens in the cell to prevent the chromosomes from segregating properly.

Every pregnancy is an independent genetic event. So although age is a factor in calculating risk of trisomy 21, Down syndrome with previous pregnancies doesn't necessarily increase a woman's risk of having another child affected by the disorder.

Some environmental factors have been implicated in Down syndrome that may increase the risk for women younger than 30. Scientists think that women who smoke while on oral contraceptives (birth control pills) may have a higher risk of decreased blood flow to their ovaries. When egg cells are starved for oxygen, they're less likely to develop normally, and nondisjunction may be more likely to occur.

Familial Down syndrome

A second form of Down syndrome, *familial Down syndrome,* is unrelated to maternal age. This disorder occurs as a result of the fusion of chromosome 21 to another autosome (often chromosome 14). This fusion is usually the result of a *translocation* — what happens when nonhomologous chromosomes exchange parts. In this case, the exchange involves the long arm of chromosome 21 and the short arm of chromosome 14. This sort of translocation is called a *Robertsonian translocation.* The leftover parts of chromosomes 14 and 21 also fuse together but are usually lost to cell division and aren't inherited. When a Robertsonian translocation occurs, affected persons can end up with several sorts of chromosome combinations in their gametes, as shown in Figure 15-4.

For familial Down syndrome, a translocation carrier has one normal copy of chromosome 21, one normal copy of chromosome 14, and one fused translocation chromosome. Carriers aren't affected by Down syndrome because their fused chromosome acts as a second copy of the normal chromosome. When a carrier's cells undergo meiosis, some of their gametes have one translocated chromosome or get the normal complement that includes one copy of each chromosome. Fertilizations of gametes with a translocated chromosome and a normal chromosome 21 produce the phenotype of Down syndrome. Roughly 10 percent of the liveborn children of carriers have trisomy 21. Carriers have a greater chance than normal of miscarriage because of monosomy (of either 21 or 14) and trisomy 14.

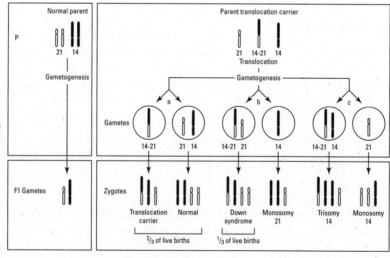

Figure 15-4: A translocation that leads to familial Down syndrome.

Other trisomies

Trisomy 18, also called *Edward syndrome,* also results from nondisjunction. About 1 in 6,000 newborns has trisomy 18, making it the second most common trisomy in humans. The disorder is characterized by severe birth defects including severe heart defects and brain abnormalities. Other defects associated with trisomy 18 include a small jaw relative to the face, clenched fingers, rigid muscles, and foot defects. Most affected infants with trisomy 18 don't live past their first birthdays. Like trisomy 21, the chance of having a baby with trisomy 18 is higher in women who become pregnant when they're older than 35.

The third most common trisomy in humans is *trisomy 13,* or *Patau syndrome.* About 1 in 12,000 live births is affected by trisomy 13; many embryos with this condition miscarry early in pregnancy. Babies born with trisomy 13 have a very short life expectancy — most die before the age of 6 months. However, some may survive until 2 or 3 years of age; records show that two children with Patau syndrome lived well into childhood (one died at 11 and the other at 19). Babies affected by trisomy 13 have extremely severe brain defects along with many facial structure defects. Absent or very small eyes and other defects of the eye, cleft lips, cleft palates, heart defects, and *polydactyly* (extra fingers and toes) are common among these children.

Another type of trisomy, *trisomy 8,* occurs very rarely (1 in 25,000 to 50,000 births). Children born with trisomy 8 have a normal life expectancy but often are affected by mental retardation and physical defects such as contracted fingers and toes.

Other things that go awry with chromosomes

In addition to monosomies and trisomies, numerous other chromosomal disorders can occur in humans. Whole sets of chromosomes can be added, or chromosomes can be broken or rearranged. This section covers some of these other sorts of chromosome disorders.

Polyploidy

Polyploidy, the occurrence of more than two sets of chromosomes, is extremely rare in humans. Two reported conditions of polyploidy are *triploid* (three full chromosome sets) and *tetraploid* (four sets). Most polyploid pregnancies result in miscarriage or stillbirth. All liveborn infants with triploidy have severe, untreatable birth defects, and most don't survive longer than a few days.

Mosaicism

Mosaicism is a form of aneuploidy that creates patches of cells with variable numbers of chromosomes. Early in embryo development, a nondisjunction similar to the one shown in Figure 15-1 can create two cells that are aneuploid (most often one cell is trisomic, with one extra chromosome copy, and the other is monosomic, with a chromosome missing its homolog). A cell can also lose a chromosome, leading to a monosomy without an accompanying trisomy. All the cells that descend from the aneuploid cells created during mitosis are also aneuploid. The magnitude of the effects of mosaicism depends on when the error occurs: If the error happens very early, most of the individual's cells are affected.

Most mosaicisms are lethal except when the mosaic cell line is confined to the placenta. Many embryos with placenta mosaics develop normally and suffer no ill effects. Sex chromosome mosaics are the most common in humans; XO-XXX and XO-XXY are common mosaic genotypes. Trisomy 21 also appears as a mosaic with normal diploid cells. Often, individuals with mosaicism are affected in the same ways as persons who are entirely aneuploid.

Fragile X syndrome

Many chromosomes have *fragile sites* — parts of the chromosome that show breaks when the cells are exposed to certain drugs or chemicals. Eighty such fragile sites are common to all humans, but other sites appear because of rare mutations. One such site, fragile X on the X chromosome, is associated with the most common inherited form of mental retardation.

Fragile X syndrome results from a mutation in a gene called *FMR1* (for Fragile Mental Retardation gene 1). Like many X-linked mutations, fragile X syndrome is recessive. Therefore, women are usually mutation carriers, and men are most often affected by the disorder. Males with fragile X syndrome usually have some form of mental retardation that can vary in severity from mild behavioral or learning disabilities all the way to severe intellectual disabilities and autism. Men and boys with fragile X syndrome often have prominent ears and long faces with large jaws.

Fragile X often shows *genetic anticipation* — that is, the disorder gets more severe from one generation to the next. Within *FMR1* is a series of three bases that are repeated over and over (see Chapter 6 for details about how DNA is put together). When the DNA is replicated (or copied; see Chapter 7), repeats can easily be added by mistake, making the repeat sequence longer. In persons with fragile X syndrome, the three bases can be repeated hundreds of times (instead of the normal 5 to 40). As the gene gets longer, the effects of the mutation become more severe, with subsequent offspring suffering stronger effects of the disorder. You can find out more about anticipation in Chapter 4.

Rearrangements

Large-scale chromosome changes are called *chromosomal rearrangements.*
Four kinds of chromosomal rearrangements, shown in Figure 15-5, are possible:

- **Duplication:** Large parts of the chromosome are copied more than once, making the chromosome substantially longer.

- **Inversion:** A section of the chromosome gets turned around, reversing the sequence of genes.

- **Deletion:** Large parts of the chromosome are lost.

- **Translocation:** Parts are exchanged between nonhomologous chromosomes.

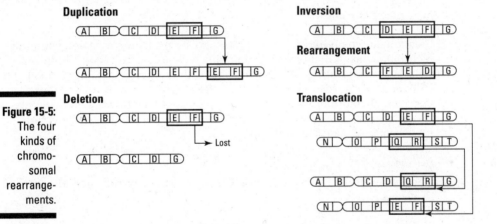

Figure 15-5: The four kinds of chromosomal rearrangements.

 All chromosomal rearrangements are mutations. Normally, mutations are very small changes within the DNA (that often have very big impacts). Mutations that involve only a few bases can't be detected by staining the chromosomes and examining the karyotype (see "What Chromosomes Reveal" for more on karyotypes). However, large-scale chromosomal changes can be diagnosed from the karyotype because they involve huge sections of the DNA. In humans, deletions and duplications are common causes of mental retardation and physical defects.

Duplications

Duplications (in this case, large, unwanted copies of portions of the chromosome) most often arise from unequal crossing-over (see "Deletions" later in this chapter). Most disorders arising from duplications are considered partial trisomies because large portions of one chromosome are usually present in triplicate.

Duplication of part of chromosome 15 is implicated in one form of autism. Autistic persons typically have severe speech impairment, don't readily interact with or respond to other persons, and exhibit ritualized and repetitive behaviors. Mental retardation may or may not be present. Persons with autism are difficult to assess because of their impaired ability to communicate. Other chromosomal rearrangements, including large-scale deletions and translocations, have also been identified in cases of autism.

Inversions

If a chromosome break occurs, sometimes DNA repair mechanisms (explained in Chapter 13) can repair the strands. If two breaks occur, part of the chromosome may be reversed before the breaks are repaired. When a large part of the chromosome is reversed and the order of the genes is changed, the event is called an *inversion*. When inversions involve the centromere, they're called *pericentric;* inversions that don't include the centromere are called *paracentric*.

Hemophilia type A may be caused, in some cases, by an inversion within the X chromosome. Patients with hemophilia have impaired blood clot formation; as a result, they bruise easily and bleed freely from even very small cuts. Mild injuries can result in extremely severe blood loss. Like most X-linked disorders, hemophilia is more common in males than females. In this case, two genes coding for the clotting factors are interrupted by the inversion, rendering both genes nonfunctional.

Deletions

Deletion, or loss, of a large section of a chromosome usually occurs in one of two ways:

- ✔ The chromosome breaks during interphase of the cell cycle (see Chapter 2 for cell cycle details), and the broken piece is lost when the cell divides.
- ✔ Parts of chromosomes are lost because of unequal crossing-over during meiosis.

Normally, when chromosomes start meiosis, they evenly align end to end with no overhanging parts. If chromosomes align incorrectly, crossing-over can create a deletion in one chromosome and an insertion of extra DNA in the other, as shown in Figure 15-6. Unequal crossover events are more likely to occur where many repeats are present in the DNA sequence (see Chapter 8 for more on DNA sequences).

Cri-du-chat syndrome is a deletion disorder caused by the loss of the short arm of chromosome 5 (varying amounts of chromosome 5 can be lost, up to 60 percent of the arm). *Cri-du-chat* is French for "cry of the cat" and refers to the characteristic, high-pitched cry that infants affected by the syndrome make. Cri-du-chat syndrome is an autosomal dominant condition; affected persons are almost always heterozygous for the mutation. Children with cri-du-chat syndrome have unusually small heads, round faces, wide-set eyes, and intellectual disabilities. Cri-du-chat syndrome is one of the more common chromosomal deletions and occurs in about 1 in 20,000 births. Most persons with cri-du-chat syndrome don't survive into adulthood. Because the majority of these deletions are new mutations, affected persons usually have no family history of cri-du-chat syndrome.

Figure 15-6:
Unequal crossover events cause large-scale deletions of chromosomes.

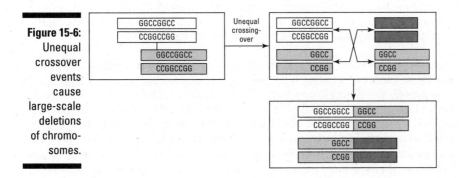

Deletion of part of the long arm of chromosome 15 results in *Prader-Willi syndrome.* This particular deletion is always in the father's chromosome, and the tendency to pass on the deletion appears to be heritable. Women with pregnancies affected by Prader-Willi syndrome usually notice that their babies start moving in the womb later and move less than unaffected babies. Affected infants are less active and have decreased muscle tone, which sometimes causes breathing problems. These infants have trouble feeding and usually don't grow at a normal rate. Children with Prader-Willi syndrome can have mental retardation, but their intellectual disabilities usually aren't very severe. Feeding problems early in life often give way to obesity later on, but persons with Prader-Willi syndrome almost always have small stature. Like cri-du-chat syndrome, Prader-Willi syndrome is often the result of spontaneous mutation (see Chapter 12 for more on genetic disorders).

Translocations

Translocations involve the exchange of large portions of chromosomes. They occur between nonhomologous chromosomes and come in two types:

- ✔ **Reciprocal translocation:** An equal (balanced) exchange in which each chromosome winds up with part of the other. This is the most common form of translocation.

- ✔ **Nonreciprocal translocation:** An uneven exchange in which one chromosome gains a section but the other chromosome does not, resulting in a deletion.

Like inversions, translocations can result from broken chromosomes that get mismatched before the repair process is complete. When two chromosomes are broken, they can exchange pieces (reciprocal or balanced translocation), gain pieces (nonreciprocal translocation), or lose pieces (deletion). When the breaks interrupt one or more genes, those genes are rendered nonfunctional.

One disorder in humans that sometimes involves a balanced translocation event is bipolar disorder. Bipolar disorder may result when chromosomes 9 and 11 exchange parts, interrupting a gene on chromosome 11. This gene, called *DIBD1* (for Disrupted in Bipolar Disorder gene 1) has also been implicated in other psychiatric disorders such as schizophrenia.

Chromosomes 11 and 22 are often involved in balanced translocation events that cause birth defects (such as cleft palate, heart defects, and mental retardation) and a hereditary form of breast cancer. Chromosome 11 seems particularly prone to breakage in an area that has many repeated sequences (where two bases, A and T, are repeated many times sequentially). Most repeated sequences like this one are considered junk DNA (see Chapter 8 for an explanation of junk DNA). Because both chromosome 11 and chromosome 22 contain similar repeat sequences, the repeats may allow crossover events to occur by mistake, resulting in a reciprocal translocation.

In many cases, a translocation event occurs spontaneously in one parent, who then passes the disrupted chromosomes on to his or her offspring, resulting in partial trisomies and partial deletions. In these cases, the carrier parents may be unaffected by the disorder.

Chapter 16

Treating Genetic Disorders with Gene Therapy

. . . .

. . . .

*T*he completion of the Human Genome Project in 2004 (see Chapter 8), along with the sequencing of nonhuman genomes, spawned an incredible revolution in the understanding of genetics. Simultaneously, geneticists raced to develop medicines to treat and cure diseases caused by genes gone awry. *Gene therapy,* treatment that gets at the direct cause of genetic disorders, is sometimes touted as the magic bullet, the cure-all for inherited diseases (see Chapter 13) and cancer (see Chapter 14). Gene therapy may even provide a way to block the genes of pathogens such as viruses, providing reliable treatments for illnesses that currently have none.

Unfortunately, the shining promise of gene therapy has been hampered by a host of challenges, including finding the right way to supply the medicine to patients without causing worse problems than the ones being treated. Moreover, the genetics of disease have turned out to be far more complicated than anyone anticipated. In this chapter, I examine the progress and perils of gene therapy.

Alleviating Genetic Disease

Take a glance through Part III of this book for proof that your health and genetics are inextricably linked. Mutations cause disorders that are passed from generation to generation, and mutations acquired during your lifetime can have unwanted consequences such as cancer. Your own genes aren't the only ones that cause complications — the genes carried by bacteria, parasites, and viruses lend a hand in spreading disease and dismay worldwide.

So wouldn't it be great if you could just turn off those pesky bad genes? Just think: A mutation causes a loss of function in a tumor suppressor gene, and you get a shot to turn that gene back on. A virus giving you trouble? Just take a pill that blocks the function of viral genes.

Some geneticists see the implementation of these genetic solutions to health problems as only a matter of time. Therefore, the development of gene therapy has focused on two major courses of action:

- ✔ Supplying genes to provide desired functions that have been lost or are missing
- ✔ Blocking genes from producing unwanted products

Finding Vehicles to Get Genes to Work

The first step in successful gene therapy is designing the right delivery system to introduce a new gene or shut down an unwanted one. The delivery system for gene therapy is called a *vector*. A perfect vector

- ✔ Must be innocuous so that the recipient's immune system doesn't reject or fight the vector.
- ✔ Must be easy to manufacture in large quantities. Just one treatment may require over 10 billion copies of the vector, because you need one delivery vehicle for every cell in the affected organ.
- ✔ Must be targeted for a specific tissue. Gene expression is tissue-specific (see Chapter 11 for details), so the vector has to be tissue-specific, too.
- ✔ Must be capable of integrating its genetic payload into each cell of the target organ so that new copies of each cell generated later by mitosis contain the gene therapy payload.

Currently, viruses are the favored vector. Most gene therapies aim to put a new gene into the patient's genome, and this gene-sharing action is almost precisely what viruses do naturally.

When a virus latches onto a cell that isn't protected from the virus, the virus hijacks all that cell's activities for the sole purpose of making more viruses. Viruses reproduce this way because they have no moving parts of their own to accomplish reproduction. Part of the virus's attack strategy involves integrating virus DNA into the host genome in order to execute viral gene expression. The problem is that when a virus is good at attacking a cell, it causes an infection that the patient's immune system fights. So the trick to using a virus as a vector is taming it.

Gentling a virus for use as a vector usually involves deleting most of its genes. These deletions effectively rob the virus of almost all its DNA, leaving only a few bits. These remaining pieces are primarily the parts the virus normally uses to get its DNA into the host. Using DNA manipulation techniques like those I describe in the "Inserting Healthy Genes into the Picture" section of this chapter, the scientist splices a healthy gene sequence into the virus to replace the deleted parts of the viral genome. But a helper is needed to move the pay-load from the virus to the recipient cell, so the scientist sets up another virus particle with some of the deleted genes from the vector. This second virus, called a *helper*, makes sure that the vector DNA replicates properly.

Geneticists conducting gene therapy have several viruses to choose from as possible delivery vehicles (vectors). These viruses fall into one of two classes:

- ✔ Those that integrate their DNA directly into the host's genome

- ✔ Those that climb into the cell nucleus to become permanent but sepa-rate residents (called *episomes*)

Within these two categories, three types of viruses — oncoretroviruses, lenti-viruses, and adenoviruses — are popular choices for gene therapy.

Viruses that join right in

Two popular viruses for gene therapy integrate their DNA directly into the host's genome. *Oncoretroviruses* and *lentiviruses* are retroviruses that trans-fer their genes into the host genome; when the retrovirus genes are in place, they're replicated right along with the other host DNA. Retroviruses use RNA instead of DNA to code their genes and use a process called *reverse transcrip-tion* (described in Chapter 11) to convert their RNA into DNA, which is then inserted into a host cell's genome.

Oncoretroviruses, the first vectors developed for gene therapy, get their name from *oncogenes,* which turn the cell cycle on permanently — one of the precursors to development of full-blown cancer. Most of the oncoret-rovirus vectors in use for gene therapy trace their history back to a virus that causes leukemia in monkeys (it's called *Moloney murine leukemia virus,* or MLV). MLV has proven an effective vector, but it's not without prob-lems; MLV's propensity to cause cancer has been difficult to keep in check. Oncoretroviruses work well as vectors only if they're used to treat cells that are actively dividing.

Lentiviruses, on the other hand, can be used to treat cells that aren't divid-ing. You're probably already familiar with a famous lentivirus: HIV. Vectors for gene therapy were developed directly from the HIV virus itself. Although the gutted virus vectors contain only 5 percent of their original DNA, rendering them harmless, lentiviruses have the potential to regain the deleted genes if they come in contact with untamed HIV virus particles (that is, the ones that infect people with AIDS). Lentiviruses are also a bit dicey because they tend to put genes right in the middle of host genes, leading to loss-of-function mutations (I detail this and other mutations in Chapter 13).

Nonetheless, HIV lentivirus vectors are used to combat AIDS. The vector virus carries a genetic message that gets stored in the patient's immune cells. When HIV attacks these immune cells, the vector DNA blocks the attacking virus from replicating itself, effectively protecting the patient from further infection. So far, this treatment seems to work, and it substantially reduces the amount of virus that affected persons carry.

Viruses that are a little standoffish

Adenoviruses are excellent vectors because they pop their genes into cells regardless of whether cell division is occurring. In gene therapy, adenoviruses have been both promising and problematic. On the one hand, they're really good at getting into host cells. On the other hand, they tend to excite a strong immune response — the patient's body senses the virus as a foreign particle and fights it. To combat the immune reaction, researchers have worked to delete the genes that make adenoviruses easy for the host to recognize.

Adenoviruses don't put their DNA directly into the host genome. Instead, they exist separately as episomes, so they aren't as likely as lentiviruses to cause mutations. The drawback is that the episomes aren't always replicated and passed on to daughter cells when the host cell divides. Nonetheless, researchers have used adenovirus vectors with notable success — and failure. (See "Progress on the Gene Therapy Front" at the end of the chapter for the details.)

Inserting Healthy Genes into the Picture

Finding the right delivery system is a necessary step in mastering gene therapy, but to nab genes and put them to work as therapists, geneticists must also find the right ones. Because finding healthy genes isn't simple, gene mapping is still a major obstacle in the road to implementing gene therapy. Imagine you're handed a man's photograph and told to find him in New York City — no name, no address, no phone number. The task of finding that man includes figuring out his identity (maybe by finding out who his friends are), figuring out what he does for a living, narrowing your search to the borough he lives in, and identifying his street, block, and finally, his address. This wild-goose chase is almost exactly like the gargantuan task of finding genes.

Your DNA has roughly 22,000 genes tucked away among about 3 billion base pairs of DNA. (Flip to Chapter 6 to find out how DNA is sized up in base pairs.) Because most genes are pretty small, relatively speaking (often less than 5,000 base pairs long; see Chapter 9), finding just one gene in the midst of all the genetic clutter may sound like a nearly impossible task. Until recently, the only tool geneticists had in the search for genes was the observation of patterns of inheritance (like those shown in Chapter 12) and the subsequent comparisons of how various groups of traits were inherited. Geneticists use this method, called *linkage analysis,* to construct gene maps (see Chapter 4). With the advent of DNA sequencing (see Chapter 8), however, the search for names and addresses of genes has reached a whole new level (but the search still isn't over; see the sidebar "The role of the Human Genome Project"). Now, geneticists hook up with an extended network of people to nail down the exact locations of genes:

- **Physicians** identify a disorder by observing a phenotype caused by mutation. Essentially, this is the face of the gene.

- **Genetic counselors** work with patients and their families to gather complete medical histories (see Chapter 12). Analysis of family trees may uncover other traits that associate with the disorder.

- **Cell biologists** look at the karyotypes of many affected people and link traits to obvious chromosomal abnormalities. These large-scale changes in chromosomes often provide hints about where genes reside. (Chapter 15 examines methods of karyotyping.)

- **Population geneticists** analyze the DNA of large groups of people with and without the disease to narrow down which chromosomes and which genes are involved with the disease.

- **Biochemists** study the chemical processes in the affected organs of people with the disease to identify the physiology of the disorder. Often, they're able to nab the precise protein gone wrong.

- **Geneticists,** with the protein in hand, use the genetic code (profiled in Chapter 10) to work backwards from the building blocks of that protein — the specific amino acids — to discern what the mRNA instructions were.

Identifying the right protein and backtracking to the mRNA pattern is extremely helpful, but it still doesn't divulge the gene's identity. Problems include the fact that mRNAs are often heavily edited before they're translated into proteins (see Chapter 10) and the fact that the code is *degenerate,* meaning that more than one codon can be used to get a particular amino acid. The protein provides a general idea of what the gene address is, but it's not precise enough. To close in on the right address, the gene hunter has to sort through the DNA itself.

The role of the Human Genome Project

Can't geneticists just look up the genes they need from all the sequencing data collected by the Human Genome Project, or HGP? Someday the answer will be yes, but we're not there yet. By 2005, 99 percent of the gene-rich part of the genome (called the *euchromatin*) was fully sequenced. That's the good news. The bad news for gene hunters is that a whopping 20 percent of the noncoding regions of the genome still aren't sequenced.

The noncoding part of the genome (the *heterochromatin*) has been tough to work with because it's made up of repetitive sequences. All that repetition makes putting the sequences into their proper order extremely difficult. For example, researchers still argue about how many total genes there are (probably around 22,000, but possibly more or fewer). And many genes are yet to be discovered; what they control and where they're located are still unknown.

Unfortunately, the maps that the HGP constructed of the entire humane genome are drawn at the wrong scale to be useful for pinpointing the locations of genes. To get an idea of how scale can be a problem, think about looking at a road map. A low-resolution highway map can help you find your way from one city to another, but it can't guide you to a very specific street address in a particular city.

What it comes down to is that geneticists continue to explore the billions of base pairs that contain the genetic instructions that make humans tick. That's why the gene hunt is likely to go on for a long time to come. (For full coverage of the HGP, flip to Chapter 8.)

The entire gene-hunting safari depends on vast computer databases that the scientific community can easily access. These databases allow investigators to search professional journals to keep up with new discoveries by other scientists. Researchers are also constantly adding new pieces of the puzzle — such as newly identified proteins — to storehouses of data.

You can take a peek into the genetic data warehouse by visiting www.ncbi. nlm.nih.gov/entrez/query.fcgi?db=OMIM. The NCBI link in the upper left part of the page leads you to the home page of the National Center for Biotechnology Information. From there, you can explore everything from DNA to protein data compiled by scientists from around the world.

Recombinant DNA technology is the catchall phrase that covers most of the methods geneticists use to examine DNA in the lab. The word *recombinant* is used because DNA from the organism being studied is often popped into a virus or bacteria (that is, it's recombined with DNA from a different source) to allow further study. Scientists also use recombinant DNA for a vast number of other applications, including creating genetically engineered organisms (see Chapter 19) and cloning (see Chapter 20). In the case of gene therapy, recombinant DNA is used to

✔ Locate the gene (or genes) involved in a particular disorder or disease

✔ Cut the desired gene out of the surrounding DNA

✔ Pop the gene into a vector (delivery vehicle) for transfer into the cells where treatment is needed

Checking out a DNA library

One of the most popular methods for tracking down a specific gene is to create a *DNA library*. That's just what it sounds like: a library filled with chunks of DNA instead of books. Geneticists can paw through the library to nail down the piece of DNA containing the gene of interest. One popular version of the genetic library method is called a *cDNA library* — a collection of genetic instruction manuals that are actually in use in a specific cell (the *c* stands for *complementary* because the whole process actually starts by copying mRNA messages into complementary DNA format).

The idea behind a cDNA library is to harvest all the mRNAs in a cell that's involved in some genetic disease (you can find out more about RNA in Chapter 8). Because gene expression is tissue-specific (see Chapter 10), the mRNAs in any given cell represent only the genes that are at work there. So instead of plowing through all 22,000 genes of the human genome to find the one that's in trouble, geneticists can narrow the target to only the few hundred that are in a particular cell.

Harvesting and converting mRNA

The first step in creating a cDNA library is harvesting mRNAs, and the fastest way to nab mRNAs is to grab their tails. When an mRNA is getting dressed for its trip out of the nucleus and into the cytoplasm for translation, a long string of adenine ribonucleotides gets hitched on the mRNA's back. This string, called a *poly-A tail*, helps protect the mRNA from decomposing before it finishes its job. To find the mRNAs that a cell's genes produce, geneticists use chemicals to break open cells, and then they strain out the mRNAs by exposing their tails to long strings of thymine nucleotides. The As (adenines) in the tails naturally hook up with the complementary Ts (thymines) because of the bases' natural affinities for each other.

Undergoing reverse transcription

After scientists harvest a cell's mRNAs, they convert the mRNAs' messages back to DNA by reversing the transcription process. *Reverse transcription* works a lot like DNA replication (see Chapter 7). The primer used for reverse transcription is a long string of Ts (thymines) complementary to the mRNA's poly-A tail. A special enzyme called *reverse transcriptase*, which is isolated from a virus, tacks dNTPs onto the primer to create a DNA copy of the mRNA.

After the DNA copy is made, the order of the bases — the As, Gs, Cs, and Ts — on the 5' end of the DNA sequence is determined (flip to Chapter 6 for how DNA's ends are numbered) using DNA sequencing (see Chapter 11). This partial DNA sequence (about 500 bases or so) is referred to as an *expressed sequence tag* (EST). It's *expressed* because only the exons are present in the DNA sequence, and *tag* comes from the fact that only part of the entire gene sequence is obtained (and therefore "tagged").

Screening the library

With ESTs created (see the preceding section), gene hunters examine every "book" in the cDNA library to find the particular gene that causes the disease. This process is called *screening* the library. The idea here is to spread out all the ESTs and sort through them to find the precise EST that came from the gene scientists are looking for. The difficulty of screening the library depends on what scientists already know about the gene. For example, knowing which protein is the one that's gone wrong can provide enough genetic information to give scientists a head start in their search. Sometimes, geneticists even look at what's known about genes with similar functions in other organisms and start there.

Regardless of the clues available to the gene hunter, screening involves making thousands of identical copies, or *clones,* of each EST by popping it into a bacteria or virus. Because ESTs are so tiny (DNA-wise), it's impossible to manipulate only one copy at a time. The cloning process separates the ESTs into neat, little, identical stacks, each composed of thousands of copies of only one EST.

One method that geneticists use to clone ESTs is called *bacteriophage cloning.* Bacteriophages (phages, for short) are handy little viruses that make a living by injecting their DNA directly into bacterial cells.

To infect bacterial cells, bacteriophages hop onto the outer cell wall and inject their DNA into the bacterium, where the phage DNA integrates directly into the bacterium's own DNA. The viral genes get replicated, transcribed, and ultimately translated using the machinery of the bacterial cells. Eventually, the phage genes set off a new phase that breaks up the bacterial DNA and frees the phage genome. The phage DNA gets replicated many times within the bacterial cells, and new phage protein shells are also produced. The bacterial cells eventually burst open, freeing the newly completed phages to infect other cells.

Here's how these funky-looking viruses get harnessed to make copies of ESTs:

1. **Geneticists take a mixture of ESTs and splice them into the DNA of thousands of bacteriophages.**

   To splice the ESTs into the phages, the phage DNA (which is circular) is cut open using a *restriction enzyme.* Restriction enzymes cut DNA at sites called *palindromes,* where the complementary sequence of bases reads the same way backwards and forwards (like 5'-GATC-3', whose complement is 3'-CTAG-5'). The restriction enzyme always cuts between the same two bases, like the G and the A, on both strands. When pulled apart, the resulting pair of cuts leaves overhanging, single-stranded ends on one long piece of phage DNA. The ESTs are treated with enzymes to give them *sticky ends* — overhanging bits complementary to the ends left in the phage DNA. When mixed together, the phage DNA and the ESTs match their sticky ends together, completing the circle of phage DNA except that each copy of the phage now contains an EST along with its own DNA.

2. **The EST-carrying phages are mixed with their favorite victims — bacteria — and poured into Petri dishes.**

3. **After the viruses spread out and do their jobs (about 24 hours after the mixing with bacteria), the result is little pits in an otherwise uniform layer of bacteria growing in the Petri dish.**

   Each little pit, called a *plaque,* represents infection caused by one phage that has reproduced and, by a chain reaction of infections, caused many bacterial cells to die and pop open. Each individual infection site represents many thousands of copies of one EST.

With thousands of ESTs and their copies, the only task that remains is finding the EST that's associated with the gene being hunted. Using the protein gone wrong as a guide, scientists can make a guess at what the EST may look like. After they decide what kind of DNA sequence may complement the EST, they order a special kit of DNA, called a *probe,* that's custom-made to match the sequence they want. The probe is complementary to all or part of the EST in question, and it's marked with dye so that scientists can find it after it bonds with the EST. Each EST is treated to make it single-stranded, and the ESTs are exposed to the probe. The probe forms a double-stranded molecule only with the EST that it matches; scientists find the matched set with special equipment that allows the dye to glow brightly.

Scientists can also use an EST to search among chromosomes to nail down the general location of a gene. The geneticist makes a *karyotype* — a collection of all the chromosomes that can be examined under the microscope (see Chapter 15). The geneticist treats the chromosomes to allow the fluorescent-dyed EST to bind with its complement on the intact chromosomes. The dyed EST sticks to the nontemplate strand from which its mRNA counterpart came. Scientists can see the results of this process with the help of a special microscope: The region where the EST attaches to its complement (the attachment process is called *hybridization*) reflects brightly under ultraviolet light. This entire procedure, called *fluorescent in situ hybridization* (or FISH, for short), allows researchers to target a region of a particular chromosome for their gene hunt, but it isn't very specific because of the way DNA is packaged (see Chapter 6). In essence, FISH narrows the target to a few million base pairs for scrutiny. But this isn't the last piece of the puzzle: With only part of the address (provided by the right EST) and the street name (the chromosome), gene hunters need to make a high-resolution map to complete their search successfully.

Mapping the gene

Thanks to the progress of the Human Genome Project, scientists have maps for each of the chromosomes, and each map has many landmarks, called *sequence tagged sites* (STSs for short). Sequence tagged sites are short stretches of unique combinations of bases scattered around the chromosome. No two STSs are alike, so they provide unique landmarks wherever they occur. A complete STS map reveals the total distance from one end of the chromosome to the other (in base pairs), as well as the landmarks along the way. Having an STS map is a bit like knowing the locations of Times Square, the Empire State Building, and Central Park relative to the entire island of Manhattan. You may know that a street you're looking for is between Central Park and the Empire State Building, but there are hundreds of little blocks to choose from in an area that size. STSs and other landmarks in the genome are a lot like that — scientists may know that an EST is between two STSs, but the STSs themselves may be 20,000 bases apart!

Using the nabbed EST as a starting point, geneticists sequence the DNA of the chromosome in both directions in a process called *chromosome walking*. Basically, they have to compile enough sequence information to run across at least two STS landmarks on the map — one in each direction. To continue with the city analogy, chromosome walking is like laying maps of neighborhoods together end to end until two major landmarks are connected. Chromosome walking provides the last two vital pieces in the puzzle: the exact location of the gene relative to the rest of the chromosome and (finally!) the entire gene sequence associated with the EST.

With new technology and knowledge of the genome, mapping genes is getting easier and easier. Projects like HapMap (covered in Chapter 17) have identified differences at the level of single nucleotides (flip to Chapter 7 to get the scoop on these DNA building blocks). These tiny differences, called *SNPs* (pronounced "snips," for *Single Nucleotide Polymorphisms*), provide such a powerful way to map genes that building a library may become unnecessary.

After researchers precisely map a gene, they compare the gene sequences of many people (both with and without a particular disease) to determine exactly what the mutation is (that is, how the gene differs between affected and unaffected people). All this information eventually winds up in the database *Online Mendelian Inheritance in Man.*

After the gene is located, many thousands of precise replicas of a healthier version of the gene can be made through a *polymerase chain reaction,* the process used for DNA fingerprinting (see Chapter 18). Researchers pop the copies of the healthy gene into the vector used for gene therapy with the same methods that they used to make the cDNA library described here.

Progress on the Gene Therapy Front

As the Human Genome Project (HGP) started fulfilling the dreams of geneticists worldwide, realizing gene therapy's promises seemed very much in reach. In fact, the first trials conducted in 1990 were a resounding success.

In those first attempts at gene therapy, two patients suffering from the same immunodeficiency disorder received infusions of cells carrying genes coding for their missing enzymes. The disorder was a form of severe combined immunodeficiency (SCID) that results from the loss of one enzyme: adenosine deaminase (ADA). SCID is so severe that affected persons must live in completely sterilized environments with no contact with the outside world, because even the slightest infection is likely to prove deadly. Because only one gene is involved, SCID is a natural candidate for treatment with gene therapy. In the HGP, retroviruses armed with a healthy *ADA* gene were infused into the two affected children with dramatic results: Both children were essentially cured of the disease and now lead normal lives.

Other implementation of gene therapies have met with mixed results. At least 17 children have been treated for an X-linked version of SCID. These children also received a retrovirus loaded with a healthy gene and were apparently cured. However, four of the children have since been diagnosed with a cancer of the blood, leukemia. The virus that delivered the gene also plopped its DNA right into a proto-oncogene, switching it on (flip to Chapter 14 for more about the actions of oncogenes).

The most famous failure of gene therapy occurred in 1999, when 18-year-old Jesse Gelsinger volunteered for a study aimed at curing a genetic disorder called ornithine transcarbamylase (OTC) deficiency. With this disorder, Jesse occasionally suffered a huge buildup of ammonia in his body because his liver lacked enough of the OTC enzyme to process all the nitrogen waste products in his blood. Jesse's disease was controlled medically — with drugs and diet — but other affected children often die of the disease. Researchers used an adenovirus to deliver a normal OTC gene directly into Jesse's liver. (See the earlier section "Viruses that are a little standoffish" for the scoop on adenoviruses.) The virus escaped into Jesse's bloodstream and accumulated in his other organs. His body went into high gear to fight what seemed like a massive infection, and four days after receiving the treatment that was meant to cure him, Jesse died. Oddly, another volunteer in the same experimental trial received the same dose of virus that Jesse did and suffered no ill effects at all.

Not all the news has been bad. In 2009, researchers announced a successful trial of gene therapy for colorblindness in monkeys. The monkeys, which had a form of red-green colorblindness similar to the sort that humans get, were given viruses bearing a functional form of the missing gene. A few weeks later, the monkeys were able to see colors that they were unable to see prior to the therapy. Also in 2009, scientists reported that by adding three genes to brains of monkeys suffering from a form of Parkinson's disease, the animals showed a decrease in the involuntary movements that accompany the disease.

Though recent results seem optimistic, the struggle to find appropriate vectors continues. The future of gene therapy is complicated by discoveries that most genetic disorders involve several genes on different chromosomes. Not only that, but many different genes can cause a given disease (diabetes, for example, is associated with genes on at least five different chromosomes), making it difficult to know which gene to treat. Finally, some genes are so large, such as the gene for Duchenne muscular dystrophy, that typical vectors can't carry them.

Part IV
Genetics and Your World

He's a mix.

In this part . . .

The technology surrounding genetics can seem bewildering, so this part aims to make understanding all the complexity less daunting.

I summarize how you can trace human history using genetics and how human activities affect the genetics of populations of animals and plants around the world. If you've ever marveled at the crime-solving power of forensics, you get all the details of DNA's contributions to the war on crime here. With the same technology used in forensics, humans can move genes from one organism to another for all sorts of reasons; I explain the perils and progress in genetic engineering and cloning in this part. And finally, because genetics knowledge opens up a lot of choices, I cover the ups and downs of ethics and genetics.

Chapter 17

Tracing Human History and the Future of the Planet

In This Chapter

▶ Relating the genetics of individuals to the genetics of groups

▶ Describing genetic diversity

▶ Understanding the genetics of evolution

*1*t's impossible to overestimate the influence of genetics on our planet. Every living thing depends on DNA for its life, and all living things, including humans, share DNA sequences. The amazing similarities between your DNA and the DNA of other living things suggest that all living things trace their history back to a single source. In a very real sense, all creatures great and small are related somehow.

You can examine the genetic underpinnings of life in all sorts of ways. One powerful method for understanding the patterns hidden in your DNA is to compare the DNA of many individuals as a group. This specialty, called *population genetics,* is a powerful tool. Geneticists use this tool to study not only human populations but also animal populations to understand how to protect endangered species, for example. By comparing DNA sequences of various species, scientists also infer how natural selection acts to create evolutionary change. In this chapter, you find out how scientists analyze genetics of populations and species to understand where we came from and where we're going.

Genetic Variation Is Everywhere

The next time you find yourself channel surfing on the TV, pause a moment on one of the channels devoted to science or animals. The diversity of life on earth is truly amazing. In fact, scientists still haven't discovered all the species living on our planet; the vast rain forests of South America, the deep-sea vents of the ocean, and even volcanoes hold undiscovered species.

The interconnectedness of all living things, from a scientific perspective, can't be overstated. The sum total of all the life on earth is referred to as *biodiversity*. Biodiversity is self-sustaining and is life itself. Together, the living things of this planet provide oxygen for you (and everything else) to breathe, carbon dioxide to keep plants alive and regulate the temperature and weather, rainwater for you and your food supply, nutrient cycling to nourish every single creature on earth, and countless other functions.

Biodiversity provides so many essential functions for human life that these services have been valued at $33 trillion a year (yes, that's trillion with a "t"). (In case you're wondering, researchers manage to put dollar values on functions that the earth performs naturally, like rainfall, oxygen production, nutrient cycles, soil formation, and pollination, to name a few.)

Underlying the world's biodiversity is *genetic variation*. When you look around at the people you know, you see enormous variation in height, hair and eye color, skin tone, body shape — you name it. That phenotypic (physical) variation implies that each person differs genetically, too. Likewise, the individuals in all populations of other sexually reproducing organisms vary in phenotype and genotype as well. Scientists describe the genetic variation in *populations* (defined as groups of interbreeding organisms that exist together in both time and space) in two ways:

- ✓ **Allele frequencies:** How often do various alleles (alternate versions of a particular section of DNA) show up in a population?

- ✓ **Genotype frequencies:** What proportion of a population has a certain genotype?

Allele frequencies and genotype frequencies are both ways of measuring the contents of the gene pool. The *gene pool* refers to all the possible alleles of all the various genes that, collectively, all the individuals of any particular species have. Genes get passed around in the form of alleles that are carried from parent to child as the result of sexual reproduction. (Of course, genes can be passed around without sex — viruses leave their genes all over the place. See Chapter 14 for one way in which viruses leave their genetic legacies.)

Allele frequencies

Alleles are various versions of a particular section of DNA (like alleles for eye color; flip to Chapter 3 for a review of terms used in genetics). Most genes have many different alleles. Geneticists use DNA sequencing (which I explain in Chapter 8) to examine genes and determine how many alleles may exist. To count alleles, they examine the DNA of many different individuals and look for differences among base pairs — the As, Gs, Ts, and Cs — that comprise DNA. For the purposes of population genetics, scientists also look for individual differences in *junk DNA* (DNA that doesn't appear to code for phenotype;

see Chapter 18 for more about how noncoding DNA is used to provide DNA fingerprints).

Some alleles are very common, and others are rare. To identify and describe patterns of commonness and rarity, population geneticists calculate allele frequencies. What geneticists want to know is what proportion of a population has a particular allele. This information can be vitally important for human health. For example, geneticists have discovered that some people carry an allele that makes them immune to HIV infection, the virus that causes AIDS.

An allele's frequency — how often the allele shows up in a population — is pretty easy to calculate: Simply divide the number of copies of a particular allele by the number of copies of all the alleles represented in the population for that particular gene.

If you know the number of *homozygotes* (individuals having two identical copies of a particular allele) and *heterozygotes* (individuals having two different alleles of a gene), you can set the problem up using these two equations: $p + q = 1$ or $q = 1 - p$. In a two-allele system, a lowercase letter p is usually used to represent one allele frequency, and q is used for the other. Always, $p + q$ must equal 1 (or 100 percent). For example, say you want to know the frequency for the dominant allele (R) for round peas in a population of plants like the ones Mendel studied (see Chapter 3 for all the details about Mendel's experiments). You know that there are 60 RR plants, 50 Rr plants, and 20 rr plants. To determine the allele frequency for R (referred to as p), you multiply the number of RR plants by 2 (because each plant has two R alleles) and add that value to the number of Rr plants: $60 \times 2 = 120 + 50 = 170$. Divide the sum, 170, by two times the total plants in the population (because each plant as two alleles), or $2(60 + 50 + 20) = 260$. The result is 0.55, meaning that 55 percent of the population of peas have the allele R. To get the frequency of r (that is, q), simply subtract 0.55 from 1.

The situation gets pretty complicated, mathematically speaking, when several alleles are present, but the take-home message of allele frequency is still the same: All allele frequencies are the proportion of the population carrying at least one copy of the allele. And all the allele frequencies in a given population must add to 1 (which can be expressed as 100 percent, if you prefer).

Genotype frequencies

Most organisms have two copies of every gene (that is, they're *diploid*). Because the two copies don't necessarily have to be identical, individuals can be either heterozygous or homozygous for any given gene. Like alleles, genotypes can vary in frequency. Genotypic frequencies tell you what proportion of individuals in a population are homozygous and, by extension, what proportion are heterozygous. Depending on how many alleles are present in a population, many different genotypes can exist. Regardless, the sum total of all the genotype frequencies for a particular locus (location on a particular chromosome; see Chapter 2 for details) must equal 1 (or 100 percent if you work in percentage instead of proportion).

To calculate a genotypic frequency, you need to know the total number of individuals who have a particular genotype. For example, suppose you're dealing with a population of 100 individuals; 25 individuals are homozygous recessive (aa), and 30 are heterozygous (Aa). The frequency of the three genotypes (assuming there are only two alleles, A and a) is shown in the following, where the total population is represented by N.

$$\text{Frequency of } AA = \frac{\text{Number of } AA \text{ individuals}}{N}$$

$$\text{Frequency of } Aa = \frac{\text{Number of } Aa \text{ individuals}}{N}$$

$$\text{Frequency of } aa = \frac{\text{Number of } aa \text{ individuals}}{N}$$

Allele frequency and genotype frequency are very closely related concepts because genotypes are derived from combinations of alleles. It's easy to see from Mendelian inheritance (see Chapter 3) and pedigree analysis (see Chapter 12) that if an allele is very common, homozygosity is going to be high. It turns out that the relationship between allele frequency and homozygosity is quite predictable. Most of the time, you can use allele frequencies to estimate genotypic frequencies using a genetic relationship called the *Hardy-Weinberg law* of population genetics, which I explain in the next section.

Breaking Down the Hardy-Weinberg Law of Population Genetics

Godfrey Hardy and Wilheim Weinberg never met, yet their names are forever linked in the annals of genetics. In 1908, both men, completely independent of each other, came up with the equation that describes how genotypic frequencies are related to allele frequencies. Their set of simple and elegant equations accurately describes the genetics of populations for most organisms. What Hardy and Weinberg realized was that in a two-allele system, all things being equal, homozygosity and heterozygosity balance out. Figure 17-1 shows how the *Hardy-Weinberg equilibrium,* as this genetic balancing act is known, looks in a graph.

Relating alleles to genotypes

An *equilibrium* occurs when something is in a state of balance. Genetically, an equilibrium means that certain values remain unchanged over the course of time. The Hardy-Weinberg law says that allele and genotype frequencies will remain unchanged, generation after generation, as long as certain conditions are met. In order for a population's genetics to follow Hardy-Weinberg's relationships:

✓ **The organism must reproduce sexually and be diploid.** Sex provides the opportunity to achieve different combinations of alleles, and the whole affair (pardon the pun) depends on having alleles in pairs (but many alleles can be used; you're not limited to two at a time).

✓ **The allele frequencies must be the same in both sexes.** If alleles depend entirely on maleness or femaleness, the relationships don't fall into place, because not all offspring have an equal chance to inherit the alleles — alleles on the Y chromosome (see Chapter 5) violate Hardy-Weinberg rules.

✓ **The loci must segregate independently.** Independent segregation of loci is the heart of Mendelian genetics, and Hardy-Weinberg is directly derived from Mendel's laws.

✓ **Mating must be random with respect to genotype.** Matings between individuals have to be random, meaning that organisms don't sort themselves out based on the genotype in question.

Figure 17-1: The Hardy-Weinberg graph describes the relationship between allele and genotype frequencies.

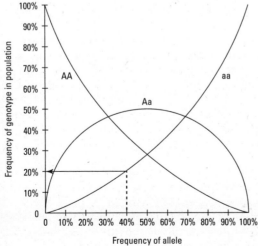

Hardy-Weinberg makes other assumptions about the populations it describes, but the relationship is pretty tolerant of violations of those expectations. Not so with the four aforementioned conditions. When one of the four major assumptions of Hardy-Weinberg isn't met, the relationship between allele frequency and genotype frequency usually starts to fall apart.

The Hardy-Weinberg equilibrium relationship is often illustrated graphically, and the Hardy-Weinberg graph is fairly easy to interpret. On the left side of the graph in Figure 17-1 is the genotypic frequency (as a percentage of the total population going from 0 at the bottom to 100 percent at the top). Across the bottom of the graph is the frequency of the recessive allele, *a* (going 0 to 100 percent, left to right). To find the relationship between genotype frequency and allele frequency according to Hardy-Weinberg, just follow a straight line up from the bottom and then read the value off the left side of the graph. For example, if you want to know what proportion of the population is homozygous *aa* when the allele frequency of *a* is 40 percent, start at the 40 percent mark along the bottom of the graph and follow a path straight up (shown in Figure 17-1 with a dashed line) until you get to the line marked *aa* (which describes the genotype frequency for *aa*). Take a horizontal path (indicated by the arrow) to the left and read the genotypic frequency. In this example, 20 percent of the population is expected to be *aa* when 40 percent of the population carries the *a* allele.

It makes sense that when the allele *a* is rare, *aa* homozygotes are rare, too. As allele *a* becomes more common, the frequency of homozygotes slowly increases. The frequency of the homozygous dominant genotype, *AA*, behaves the same way but as a mirror image of *aa* (the genotype frequency for *aa*) because of the relationship of the alleles *A* to *a* in terms of their own frequency: $p + q$ must equal 1. If p is large, q must be small and vice versa.

Check out the humped line in the middle of Figure 17-1. This is the frequency of heterozygotes, *Aa*. The highest proportion of the population that can be heterozygous is 50 percent. You may guess that's the case just by playing around with monohybrid crosses like those I describe in Chapter 3. No matter what combination of matings you try (*AA* with *aa*, *Aa* with *Aa*, *Aa* with *aa*, and so on), the largest proportion of *Aa* offspring you can ever get is 50 percent. Thus, when 50 percent of the population is heterozygous, the Hardy-Weinberg equilibrium predicts that 25 percent of the population will be homozygous for the *A* allele, and 25 percent will be homozygous for the *a* allele. This situation occurs only when p is equal to q — in other words, p equals q equals 50 percent.

Many loci obey the rules of the Hardy-Weinberg law in spite of the fact that the assumptions required for the relationship aren't met. One of the major assumptions that's often violated among humans is random mating. People tend to marry each other based on their similarities, such as religious background, skin color, and ethnic characteristics. For example, people of similar socioeconomic background tend to marry each other more often than chance would predict. Nevertheless, many human genes are still in Hardy-Weinberg equilibrium. That's because matings may be dependent on some characteristics but are still independent with respect to the genes. The gene that confers immunity to HIV is a good example of a locus in humans that obeys the Hardy-Weinberg law despite the fact that its frequency was shaped by a deadly disease.

Violating the law

Populations can wind up out of Hardy-Weinberg equilibrium in several ways. One of the most common departures from Hardy-Weinberg occurs as a result of *inbreeding*. Put simply, inbreeding happens when closely related individuals mate and produce offspring. Purebred dog owners are often faced with this problem because certain male dogs sire many puppies, and a generation or two later, descendents of the same male are mated to each other. (In fact, selective inbreeding is what created the various dog breeds to begin with.)

Inbreeding tends to foul up Hardy-Weinberg because some alleles start to show up more and more often than others. In addition, homozygotes get more common, meaning fewer and fewer heterozygotes are produced. Ultimately, the appearance of recessive phenotypes becomes more likely. For example, the appearance of hereditary problems in some breeds of animals, such as deafness among Dalmatian dogs, is a result of generations of inbreeding.

The high incidence of particular genetic disorders among certain groups of people, such as Amish communities (see Chapter 12), is also a result of inbreeding. Even if people in a group aren't all that closely related anymore, if a small number of people started the group, everyone in the group is related somehow. (Relatedness shows up genetically even within large human populations; take a look at the section "Mapping the Gene Pool" in this chapter to find out how.)

Loss of heterozygosity is thought to signal a population in peril. Populations with low levels of heterozygosity are more vulnerable to disease and stress, and that vulnerability increases the probability of extinction. Much of what's known about loss of heterozygosity and resulting problems with health of individuals — a situation ironically called *inbreeding depression* — comes from observations of captive animals, like those in zoos. Many animals in zoos are descended from captive populations, populations that had very few founders to begin with. For example, all captive snow leopards are reportedly descended from a mere seven animals.

Not just captive animals are at risk. As habitats for animals become more and more altered by human activity, natural populations get chopped up, isolated, and dwindle in size. Conservation geneticists, like yours truly, work to understand how human activities affect natural populations of birds and animals. See the sidebar "Genetics and the modern ark" for more about how zoos and conservation geneticists work to protect animals from inbreeding depression and rescue species from extinction.

CASE STUDY

Genetics and the modern ark

As human populations grow and expand, natural populations of plants and animals start getting squeezed out of the picture. One of the greatest challenges of modern biology is figuring out a way to secure the fate of worldwide biodiversity. Preserving biodiversity often takes two routes: establishment of protected areas and captive breeding.

Protected areas such as parks set aside areas of land or sea to protect all the creatures (animals and plants) that reside within its borders. Some of the finest examples of such efforts are found among America's national parks. But although protecting special areas helps preserve biodiversity, these islands of biodiversity also allow populations to become isolated. With isolation, smaller populations start to inbreed, resulting in genetic disease and vulnerability to extinction. Sometimes, it's necessary for conservation geneticists to step in and lend a hand to rescue these isolation populations from genetic peril. For example, greater prairie chickens were common in the Midwest at one time. By 1990, their populations were tiny and isolated. Isolation contributed to inbreeding, causing their eggs to fail to hatch. In order

to help rebuild a healthy population, biologists brought in more birds from populations elsewhere to increase genetic diversity. The strategy worked — the prairie chickens' eggs now hatch with healthy chicks that are hoped to bring the population back from the brink of extinction.

Captive breeding efforts by zoos, wildlife parks, and botanical gardens are also credited with preserving species. Twenty-five animal species that are completely extinct in the wild still survive in zoos thanks to captive breeding programs. Most programs are designed to provide not only insurance against extinction but also breeding stock for eventual reintroduction into the wild. Unfortunately, zoo populations often descend from very small founder populations, causing considerable problems with inbreeding. Inbreeding leads to fertility problems and the death of offspring shortly after birth. In the last 20 years, zoos and similar facilities have worked to combat inbreeding by keeping track of pedigrees (like the ones that appear in Chapter 12) and swapping animals around to minimize sexual contact between related animals.

Mapping the Gene Pool

When the exchange of alleles, or *gene flow,* between groups is limited, populations take on unique genetic signatures. In general, unique alleles are created through mutations (see Chapter 13). If groups of organisms are geographically separated and rarely exchange mates, mutant alleles become common within populations. What this amounts to is that some alleles are found in only certain groups, giving each group a unique genetic identity. (After some time, these alleles usually conform to a Hardy-Weinberg equilibrium within each population; see the section "Breaking Down the Hardy-Weinberg Law of Population Genetics" for details.) Geneticists identify genetic signatures of unique alleles by looking for distinct patterns within genes and certain sections of junk DNA (see Chapter 18 for how junk DNA conceals genetic information).

Mutant alleles that show up outside the population they're usually associated with suggest that one or more individuals have moved or dispersed between populations. Geneticists use these genetic hints to trace the movements of animals, plants, and even people around the world. In the sections that follow, I cover some of the latest efforts to do just that.

One big happy family

With the contributions of the Human Genome Project (covered in Chapter 11), human population geneticists have a treasure trove of information to sift through. Using new technologies, researchers are learning more than ever before about what makes various human populations distinct. One such effort is the HapMap Project. Hap stands for *haplotype,* which is another way of saying an inventory of human alleles. The alleles being studied for the HapMap aren't necessarily alleles from specific genes; many are alleles within the junk DNA. The HapMap takes advantage of single base pair changes, called SNPs (see Chapter 18), in the DNA; SNPs are the results of thousands of substitution mutations. Most of these tiny changes have no effect on phenotype, but collectively they vary enough from one population to another to allow geneticists to discern each population's genetic signature.

After geneticists understand how much diversity exists among haplotypes, they work to create genetic maps that relate SNP alleles to geographic locations. Essentially, all humans tend to divide up genetically into the three continents of Africa, Asia, and Europe. This isn't too surprising — humans have been in North and South America for only 10,000 years or so. When the genetic uniqueness of the Old World's people was described, geneticists examined populations in North America and other immigrant populations to see if genetics could predict where people came from. For example, genetic analyses of a group of immigrants in Los Angeles accurately determined which continent these people originally lived on. Some geneticists believe that the genetic maps can be even more specific and may point people to countries, and maybe even cities, where their ancestors once lived. The ultimate goal of the HapMap Project is to link haplotypes to populations along with information about the environment, family histories, and medical conditions to development tailor-made treatments for diseases.

Because humans love to travel, geneticists have also compared rates of movements between men and women. Common wisdom suggests that, historically, men tended to move around more than women did (think Christopher

Columbus or Leif Ericson). However, DNA evidence suggests that men aren't as prone to wander as previously believed. Geneticists compared mitochondrial DNA (passed from mother to child) with Y chromosome DNA (passed from father to son). It seems that women have migrated from one continent to another eight times more frequently than males. The tradition of women leaving their own families to join their husbands may have contributed to the pattern, but another possible explanation exists: A pattern of *polygyny,* men fathering children by more than one woman. So, back to that bit about men wandering. . . .

Uncovering the secret social lives of animals

Gene flow can have an enormous impact on threatened and endangered species. For example, scientists in Scandinavia were studying an isolated population of gray wolves not long ago. Genetically, the population was very inbred; all the animals descended from the same pair of wolves. Heterozygosity was low and, as a consequence, so were birth rates. When the population suddenly started to grow, the scientists were shocked. Apparently, a male wolf migrated over 500 miles to join the pack and father wolf pups. Just one animal brought enough new genes to rescue the population from extinction.

Mating patterns of animals often provide biologists with surprises. Because humans like to form monogamous pairs, scientists have compared birds to humankind by pointing to our apparently similar mating habits. As it turns out, birds aren't so monogamous after all. In most species of perching birds (the group that includes pigeons and sparrows, to name two widespread types), 20 percent of all offspring are fathered by some male other than the one with whom the female spends all her time. By spreading paternity among several males, a female bird makes sure that her offspring are genetically diverse. And genetic diversity is incredibly important to help fend off stress and disease.

Genetics reveals that some birds are really frisky. For example, fairy wrens — tiny, brilliant-blue songbirds — live in Australia in big groups; one female is attended by several males who help her raise her young. But none of the males attending the nest actually fathers any of the kids — female fairy wrens slip off to mate with males in distant territories. Other birds form family groups. Florida scrub jays — beautiful aquamarine natives of central Florida — stay home and help mom and dad raise younger brothers and sisters. Eventually, older kids inherit their parents' territory. Another Australian species, white-winged choughs, put a whole different twist on gathering a labor force for raising their kids. Chough (pronounced *chuff*) families kidnap their neighbors' kids and put them to work raising offspring.

It turns out that humans aren't the only ones who live in close association with their parents, brothers, or sisters for their entire lives. Some species of whales live in groups called *pods*. Every pod represents one family: moms, sisters, brothers, aunts, and cousins, but not dads. Different pods meet up to find mates — as in the son/brother of one pod may mate with the daughter/sister of another pod. Males father offspring in different pods but stay with their own families for their entire lives. Sadly, geneticists learned about whale family structures and mating habits by taking meat from whales that had been killed by people. Like so many of the world's creatures, whales are killed by hunters. Hopefully, though, the information that scientists gather when whales are harvested will contribute to their conservation, allowing the planet's amazing biodiversity to persist for generations to come.

Changing Forms over Time: The Genetics of Evolution

Evolution, or how organisms change over time, is a foundational principle of biology. When Charles Darwin put forth his observations about natural selection, the genetic basis for inheritance was unknown. Now, with powerful tools like DNA sequencing (which appears in Chapter 8), scientists are documenting evolutionary change in real time, as well as uncovering how species share ancestors from long ago.

When genetic variation arises (from mutation, which I talk about in Chapter 13), new alleles are created. Then, *natural selection* acts to make particular genetic variants more common by way of improved survival and reproductive success for some individuals over others. In this section, you discover how genetics and evolution are inextricably tied together.

Genetic variation is key

All evolutionary change occurs because genetic variation arises through mutation. Without genetic variation, evolution can't take place. While many mutations are decidedly bad (I discuss those in Chapter 13), some mutations confer an advantage, such as resistance to disease.

No matter how a mutation arises or what consequences it causes, the change must be heritable, or passed from parent to offspring, to drive evolution.

Until recently, it wasn't possible to examine heritable variation directly. Instead, phenotypic variation was used as an indicator of how much genetic variation might exist. With the help of DNA sequencing, scientists have come to realize that genetic variation is vastly more complex than anyone ever imagined.

Heritable genetic variation alone doesn't mean that evolution will occur, however. The final piece in the evolutionary puzzle is natural selection. Put simply, natural selection occurs when conditions favor individuals carrying particular traits. By favor, it's meant that those individuals reproduce and survive better than other individuals carrying a different set of traits. This success is sometimes referred to as *fitness,* which is the degree of reproductive success associated with a particular genotype. When an organism has high fitness, its genes are being passed on successfully to the next generation. Through its effects on fitness, natural selection produces *adaptations,* or sets of traits that are important for survival. The white fur of polar bears, which allow them to blend into the snowy landscape of Arctic regions, is an example of an adaptation.

Where new species come from

Probably since the dawn of time (or at least the dawn of humankind, anyhow), humans have been classifying and naming the creatures around them. The formalized species naming system, what scientists call *taxonomic classification,* has long relied on physical differences and similarities between organisms as a means of sorting things out. For example, elephants from Asia and elephants from Africa are obviously both elephants, but they're so different in their physical characteristics, among other things, that they're considered separate species. Over the past 50 years or so, the way in which species are classified has changed as scientists have gained more genetic information about various organisms.

One way of classifying species is the *biological species concept,* which bases its classification on reproductive compatibility. Organisms that can successfully reproduce together are considered to be of the same species, and those that can't reproduce together are a different species. This definition leaves a lot to be desired, because many closely related organisms can interbreed yet are clearly different enough to be separate species.

Another method of classification, one that works a bit better, says that species are groups of organisms that maintain unique identities — genetically, physically, and geographically — over time and space. A good example of this definition of species is dogs and wolves. Both dogs and wolves are in the same pigeonhole, so to speak — they're both in the genus *Canis.* (Sharing a genus name tells you that organisms are quite similar and very closely related.) But their species names are different. Dogs are always *Canis familiaris,* but there are many species of wolves, all beginning with *Canis* but ending with a variety of species names to accurately describe how different they are from each other (such as gray wolves, *Canis lupus,* and red wolves, *Canis rufus*). Genetically, dogs and wolves are very distinct, but they aren't so different that they can't interbreed. Dogs and wolves occasionally mate and produce offspring, but left to their own devices, they don't interbreed.

When populations of organisms become reproductively isolated from each other (that is, they no longer interbreed), each population begins to evolve independently. Different mutations arise, and with natural selection, the passage of time leads to the accumulation of different adaptations. In this way, after many generations, populations may become different species.

A famous example of this sort of evolutionary change comes from the aptly named Darwin's finches, a group of birds found on the Galapagos, a cluster of islands off the coast of South America. Genetic studies indicate that all Darwin's finch species are descended from a single ancestral species that landed on the islands between two and three million years ago. As islands appeared and disappeared because of volcanic activity, the birds moved from one island to another and their populations became isolated, allowing evolutionary changes and natural selection to mold each species in different ways. Thus, some Darwin's finch species have huge bills adapted to cracking open hard seeds, whereas others have dainty, slender bills for probing crevices to catch insects. Figure 17-2 gives you an idea of the diversity and relationships among these fascinating finches.

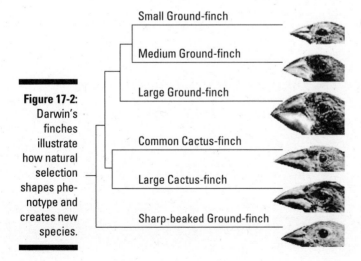

Figure 17-2:
Darwin's finches illustrate how natural selection shapes phenotype and creates new species.

Small Ground-finch

Medium Ground-finch

Large Ground-finch

Common Cactus-finch

Large Cactus-finch

Sharp-beaked Ground-finch

Growing the evolutionary tree

One of the basic concepts behind evolution is that organisms have similarities because they're related by descent from a common ancestor. Genetics and DNA sequencing techniques have allowed scientists to study these evolutionary relationships, or *phylogenies,* among organisms. For example, the DNA sequence of a particular gene may be compared across many organisms. If the gene is very similar or unchanged from one species to another, the species would be considered more closely related (in an evolutionary sense) than species that have accumulated many mutational changes in the same gene.

One way to represent the evolutionary relationships is with a tree diagram. In a similar fashion to the pedigrees used to study genetics in family relationships (flip to Chapter 12 for pedigree analysis), evolutionary trees like the one in Figure 17-2 illustrate the family relationships among species. The trunk of an evolutionary tree represents the common ancestor from which all other organisms in the tree descended. The branches of the tree show the evolutionary connections between species. In general, shorter branches indicate that species are more closely related.

Chapter 18

Solving Mysteries Using DNA

• •

In This Chapter

▶ Generating DNA fingerprints

▶ Using DNA to match criminals to crimes

▶ Identifying persons using DNA from family members

• •

*F*orensics pops up in every cop drama and murder mystery on television these days, but what is forensics used for in the real world? Generally, *forensics* is thought of as the science used to capture and convict criminals; it includes everything from determining the source of carpet fibers and hairs to paternity testing. Technically, forensics is the application of scientific methods for legal purposes. Thus, *forensic genetics* is the exploration of DNA evidence — who is it, who did it, and who's your daddy.

Just as each person has his or her own unique fingerprint, every human (with the exception of identical twins) is genetically unique. *DNA fingerprinting,* also known as *DNA profiling,* is the process of uncovering the patterns within DNA. DNA fingerprinting is at the heart of forensic genetics and is often used to

✔ Confirm that a person was present at a particular location

✔ Determine identity (including gender)

✔ Assign paternity

In this chapter, you step inside the DNA lab to discover how scientists solve forensic mysteries by identifying individuals and family relationships using genetics.

 The knowledge that every human's fingerprints are unique is probably as old as humanity itself. But Edward Henry was the first police officer to apply the patterns of loops, arches, and whorls from people's fingertips to identify individual people and match criminal to crime way back in 1899.

Rooting through Your Junk DNA to Find Your Identity

It's obvious just from looking at the people around you that each one of us is unique. But getting at the *genotype* (genetic traits) behind the *phenotype* (physical traits) is tricky business, because almost all your DNA is exactly like every other human's DNA. Much of what your DNA does is provide the information to run all your body functions, and most of those functions are exactly the same from one human to another. If you were to compare your roughly 3 billion base pairs of DNA (see Chapter 6 for how DNA is put together) with your next-door neighbor's DNA, you'd find that 99.999 percent of your DNA is exactly the same.

So what makes you look so different from the guy next door, or even from your mom and dad? Your genetic uniqueness is the result of sexual reproduction. (For more on how sexual reproduction works to make you unique, turn to Chapter 2.) Until the human genome was sequenced (see Chapter 8), the tiny differences produced by recombination and meiosis that make you genetically unique were very hard to isolate. But in 1985, a team of scientists in Britain figured out how to profile a tiny bit of DNA uniqueness into a DNA fingerprint. Surprisingly, DNA fingerprinting doesn't use the information contained in your genes that make you look unique. Instead, the process takes advantage of part of the genome that doesn't seem to do anything at all: junk DNA.

Less than 2 percent of the human genome codes for actual physical traits — that is, all your body parts and the ways they function. That's pretty astounding considering that your genome is so huge. So what's all that extra DNA doing in there? Scientists are still trying to figure that part out, but what they do know is that some junk DNA is very useful for identifying individual people.

Even within junk DNA, one human looks much like another. But short stretches of junk DNA vary a lot from person to person. *Short tandem repeats* (STRs) are sections of DNA arranged in back-to-back repetition (a simple sequence is repeated several times in a row). A naturally occurring junk DNA sequence may look something like the following example. (The spaces in these examples allow you to read the sequences more easily. Real DNA doesn't have spaces between the bases.)

TGCT AGTC AAAG TCTT CGGT TCAT

A short STR may look like this:

TCAT TCAT TCAT TCAT TCAT TCAT

The number of repeats in pairs of STRs varies from one person to another. The variations are referred to as *alleles* (see Chapter 3 for more about alleles). Using two chromosome pairs from different suspects, Figure 18-1 shows how the same STRs can have different alleles. Chromosome 1 has two loci (in reality, this chromosome may have hundreds of loci, but we're only looking at two in this example). For the first suspect, the marker at STR Locus A is the same length on both chromosomes, meaning that Suspect One is homozygous for Locus A (*homozygous* means the two alleles at a particular locus are identical). At Locus B, Suspect One (S1) has alleles that are different lengths, meaning he's heterozygous at that locus (*heterozygous* means the two alleles differ). Now look at the STR DNA profile of Suspect Two (S2). It shows the same two loci, but the patterns are different. At Locus A, S2 is heterozygous and has one allele that's different from S1's. At Locus B, S2 is homozygous for a completely different allele than the ones S1 carries.

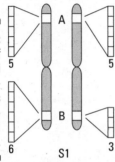
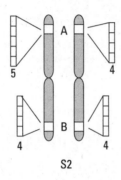

Figure 18-1: Alleles of two STR loci on the chromosomes of two suspects (S1 and S2).

The variation in STR alleles is called *polymorphism* (*poly-* meaning "many" and *morph-* meaning "shape or type"). Polymorphism arises from mistakes made during the DNA copying process (called *replication;* see Chapter 7). Normally, DNA is copied mistake-free during replication. But when the enzyme copying the DNA reaches an STR, it often gets confused by all the repeats and winds up leaving out one repeat — like one of the TCATs in the previous example — or putting an extra one in by mistake. As a result, the sequences of DNA before and after the STR are exactly the same from one person to the next, but the number of repeats *within* the STR varies. In the case of junk DNA, the mutations in the STRs create lots of variation in how many repeats appear (variations, or mutations, in genes can produce harmful and severe consequences; see Chapter 13 for details).

The specific STRs used in forensics are referred to as *loci* or *markers*. Loci is the plural form of *locus,* which is Latin for "place." (Genes are also referred to as loci; see Chapter 3.) You have hundreds of STR markers on every one of your chromosomes. These loci are named using numbers and letters, such as D5S818 or VWA.

Your STR DNA fingerprint is completely different from everyone else's on earth (except your identical twin, if you have one). The probability of anyone else having identical alleles as you at every one of his or her loci is outrageously small.

Lots of other organisms have STR DNA and, in turn, their own DNA fingerprints. Dogs, cats, horses, fish, plants — in fact, just about all eukaryotes have lots of STR DNA. (*Eukaryotes* are organisms whose cells have nuclei.) That makes STR DNA fingerprinting an extremely powerful tool to answer all kinds of biological questions. (See Chapter 17 for more about how DNA fingerprinting is used to solve other kinds of biological mysteries.)

Investigating the Scene: Where's the DNA?

When a crime occurs, the forensic geneticist and the crime scene investigator are interested in the biological evidence, because cells in biological evidence contain DNA. Biological evidence includes blood, saliva, semen, and hair.

Pets and plants play detective

DNA evidence from almost any source may provide a link between criminal and crime. For example, a particularly brutal murder in Seattle was solved entirely on the basis of DNA provided by the victim's dog. After two people and their dog were shot in their home, two suspects were arrested in the case, and blood-spattered clothing was found in their possession. The only blood on the clothing was of canine origin, and the dog's blood turned out to be the only evidence linking the suspects to the crime scene. Using markers originally designed for canine paternity analysis, investigators generated a DNA fingerprint from the dog's blood and compared it with DNA tests from the bloodstained clothing. A perfect match resulted in a conviction.

Practically any sort of biological material can provide enough DNA to match a suspect to a crime. In one murder case, the perpetrator stepped in a pile of dog feces near the scene. DNA fingerprinting matched the evidence on a suspect's shoe to the evidence at the scene, leading to a conviction. In another case, a rape victim's dog urinated on the attacker's vehicle, allowing investigators to match the pup to the truck; the suspect promptly confessed his guilt.

Even plants have a space in the DNA evidence game. The very first time DNA evidence from plants was used was in an Arizona court case in 1992. A murder victim was found near a desert tree called Paloverde. Seeds from that type of tree were found in the bed of a pickup truck that belonged to a suspect in the case, but the suspect denied ever having been in the area. The seeds in the truck were matched to the exact tree where the victim was found using DNA fingerprinting. The seeds couldn't prove the suspect's presence, but they provided a link between his truck and the tree where the body was found. The DNA evidence was convincing enough to obtain a conviction in the case.

Collecting biological evidence

Anything that started out as part of a living thing may provide useful DNA for analysis. In addition to human biological evidence (blood, saliva, semen, and hair), plant parts like seeds, leaves, and pollen, as well as hair and blood from pets, can help link a suspect and victim. (See the sidebar "Pets and plants play detective" for more on how nonhuman DNA is used to investigate crimes.)

To properly collect evidence for DNA testing, the investigator must be very, very careful, because his or her own DNA can get mixed up with DNA from the scene. Investigators wear gloves, avoid sneezing and coughing, and cover their hair (I'm not kidding — dandruff has DNA, too).

To conduct a thorough investigation, the investigator needs to collect everything at the scene (or from the suspect) that may provide evidence. DNA has been gathered from bones, teeth, hair, urine, feces, chewing gum, cigarette butts, toothbrushes, and even earwax! Blood is the most powerful evidence because even the tiniest drop of blood contains about 80,000 white blood cells, and the nucleus of every white blood cell contains a copy of the donor's entire genome and more than enough information to determine identity using a DNA fingerprint. But even one skin cell has enough DNA to make a fingerprint (see "Outlining the powerful PCR process"). That means that skin cells clinging to a cigarette butt or an envelope flap may provide the evidence needed to place a suspect at the scene.

To draw information and conclusions from the DNA evidence, the investigator needs to collect samples from the victim or victims, suspects, and witnesses for comparison. Investigators collect samples from houseplants, pets, or other living things nearby to compare those DNA fingerprints to the DNA evidence. After the investigator gets these samples, it's time to head to the lab.

Decomposing DNA

DNA, like all biological molecules, can decompose; that process is called *degradation*. *Exonucleases*, a particular class of enzymes whose sole function is to carry out the process of DNA degradation, are practically everywhere: on your skin, on the surfaces you touch, and in bacteria. Anytime DNA is exposed to exonuclease attack, its quality rapidly deteriorates because the DNA molecule starts to get broken into smaller and smaller pieces. Degradation is bad news for evidence because

DNA begins to degrade as soon as cells (like skin or blood) are separated from the living organism. To prevent DNA evidence from further degradation after it's collected, it's stored in a sterile (that is, bacteria-free) container and kept dry. As long as the sample isn't exposed to high temperatures, moisture, or strong light, DNA evidence can remain usable for more than 100 years. (Even under adverse conditions, DNA can sometimes last for centuries, as I explain in Chapter 6.)

Moving to the lab

Biological samples contain lots of substances besides DNA. Therefore, when an investigator gets evidence to the lab, the first thing to do is extract the DNA from the sample. (For a DNA extraction experiment using a strawberry, see Chapter 6.) There are different methods to extract DNA, but they generally follow these three basic steps:

1. Break open the cells to free the DNA from the nucleus (this is called *cell lysis*).

2. Remove the proteins (which make up most of the biological sample) by digesting them with an enzyme.

3. Remove the DNA from the solution by adding alcohol.

After the DNA from the sample is isolated, it's analyzed using a process called the *polymerase chain reaction,* or PCR.

Outlining the powerful PCR process

The goal of the PCR process is to make thousands of copies of specific parts of the DNA molecule — in the case of forensic genetics, several target STR loci that are used to construct a DNA fingerprint. (Copying the entire DNA molecule would be useless because the uniqueness of each individual person is hidden among all that DNA.) Many copies of several target sequences are necessary, for two reasons:

✔ Current technology used in DNA fingerprinting can't detect the DNA unless large amounts are present, and to get large amounts of DNA, you have to make copies.

✔ Matches must be exact when it comes to DNA fingerprinting and forensic genetics; after all, people's lives are on the line. To avoid misidentifications, many STR loci from each sample must be examined.

In the U.S., 13 standard markers are used for matching human samples, plus one additional marker that allows determination of gender (that is, whether the sample came from a male or a female). These markers are part of CODIS, the COmbined DNA Index System, which is the U.S. database of DNA fingerprints.

Here's how PCR works, as shown in Figure 18-2:

1. To replicate DNA using PCR, you have to separate the double-stranded DNA molecule (called the *template*) into single strands. This process is called *denaturing.* When DNA is double-stranded, the bases are

protected by the phosphate sugar backbone of the double helix (see Chapter 6). DNA's complementary bases, where all the information is stored, are locked away, so to speak. To pick the lock, get at the code, and build a DNA copy, the double helix must be opened up. The hydrogen bonds that hold the two DNA strands together are very strong, but they can be broken by heating the molecule up to a temperature just short of boiling (212 degrees Fahrenheit). When heated, the two strands slowly come apart as the hydrogen bonds melt. DNA's sugar-phosphate backbone isn't damaged by heat, so the single strands stay together with the bases still in their original order.

2. When denaturing is complete, the mix is cooled slightly. Cooling allows small, complementary pieces of DNA called *primers* to attach themselves to the template DNA. The primers match up with their complements on the template strands in a process called *annealing.* Primers only attach to the template strand when the match is perfect; if no exact match is found, the next step in the PCR process doesn't occur because primers are required to start the copying process (see Chapter 7 for more on why primers are necessary to build strands of DNA from scratch). The primers used in PCR are marked with dyes that glow when exposed to the right wavelength of light (think fluorescent paint under a black light). STRs of similar length (even though they may actually be on entirely different chromosomes, as in Figure 18-1) are labeled with different colors so that when the fingerprint is read, each locus shows up as a different color (see "Constructing the DNA fingerprint" later in this chapter).

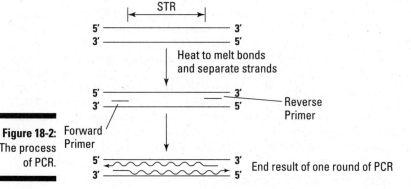

Figure 18-2: The process of PCR.

3. After the primers find their matches on the template strands, *Taq polymerase* begins to do its work. Polymerases act to put things together. In this case, the thing getting put together is a DNA molecule.

Taq polymerase starts adding bases — this stage is called *extension* — onto the 3' ends of the primers by reading the template DNA strand to determine which base belongs next (see Chapter 6 for details on DNA strands' numbered ends). Meanwhile, on the opposite template strand at the end of the reverse primer, Taq rapidly adds complementary bases using the template as a guide. (The newly replicated DNA remains double-stranded throughout this process because the mixture isn't hot enough to melt the newly formed hydrogen bonds between the complementary bases.)

One complete round of PCR produces two identical copies of the desired STR. But two copies aren't enough to be detected by the lasers used to read the DNA fingerprints (see "Constructing the DNA fingerprint"). You need hundreds of thousands of copies of each STR, so the PCR process — denaturing, annealing, and extending — repeats over and over.

Figure 18-3 shows you how fast this copying reaction adds up — after 5 cycles, you have 32 copies of the STR. Typically, a PCR reaction is repeated for 30 cycles, so with just one template strand of DNA, you end up with 1,073,741,824 copies of the target STR (the primers and the sequence between them). Usually, evidence samples consist of more than one cell, so it's likely that you start with 80,000 or so template strands instead of just one. With 30 rounds of PCR, this would yield . . . I'll wait while you do the math . . . okay, so it's a lot of DNA, as in trillions of copies of the target STR. That's the power of PCR. Even the tiniest drop of blood or a single hair can yield a fingerprint that may free the innocent or convict the guilty.

The invention of PCR revolutionized the study of DNA. Basically, PCR is like a copier for DNA but with one big difference: A photocopier makes facsimiles; PCR makes real DNA. Before PCR came along, scientists needed large amounts of DNA directly from the evidence to make a DNA fingerprint. But DNA evidence is often found and collected in very tiny amounts. Often, the evidence that links a criminal to a crime scene is the DNA contained in a single hair! One of the biggest advantages to PCR is that a very tiny amount of DNA — even one cell's worth! — can be used to generate many exact copies of the STRs used to create a DNA fingerprint (see the section "Rooting through Your Junk DNA to Find Your Identity" earlier in this chapter for a full explanation of STR). Chapter 22 looks at the discovery of PCR in more detail.

Constructing the DNA fingerprint

For each DNA sample taken as forensic evidence and put through the process of PCR, several loci are examined. ("Several" often means 13 because of the CODIS database; see the earlier "Outlining the powerful PCR process" section.) This study yields a unique pattern of colors and sizes of STRs — this is the DNA fingerprint of the individual from whom the sample came.

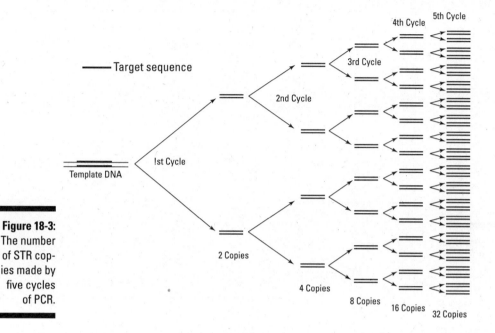

Figure 18-3:
The number of STR copies made by five cycles of PCR.

DNA fingerprints are "read" using a process called *electrophoresis,* which takes advantage of the fact that DNA is negatively charged. An electrical current is passed through a jelly-like substance (such as a gel), and the completed PCR is injected into the gel. By electrical attraction, the DNA moves toward the positive pole (electrophoresis). Small STR fragments move faster than larger ones, so the STRs sort themselves according to size (see Figure 18-4). Because the fragments are tagged with dye, a computer-driven machine with a laser is used to "see" the fragments by their colors. The STR fragments show up as peaks like those shown in Figure 18-4. The results are stored in the computer for later analysis of the resulting pattern.

The technology used in DNA fingerprinting now allows the entire process, from extracting the DNA to reading the fingerprint, to be done very rapidly. If everything goes right, it takes less than 24 hours to generate one complete DNA fingerprint.

The first cases to use DNA fingerprinting appeared in the courts in 1986. Generally, legal evidence must adhere to what experts call the *Frye standard. Frye* is short for *Frye v. United States,* a court case decided in 1923. Put simply, Frye says that scientific evidence can only be used when most scientists agree that the methods and theory used to generate the evidence are well established. Following its development, DNA fingerprinting rapidly gained acceptance by the courts, and it's now considered routine. The tests used to generate DNA fingerprints have changed over the years, and STRs are now the gold standard.

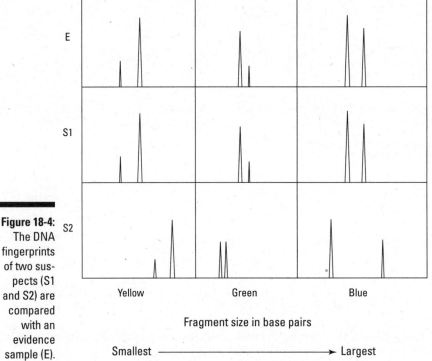

Figure 18-4: The DNA fingerprints of two suspects (S1 and S2) are compared with an evidence sample (E).

Yellow   Green   Blue

Fragment size in base pairs

Smallest ────────────────────→ Largest

Employing DNA to Catch Criminals (And Free the Innocent)

After the forensic geneticist generates DNA fingerprints from different samples, the next step is to compare the results. When it comes to getting the most information out of the fingerprints, the basic idea is to look for matches between the

✔ Suspect's DNA and DNA on the victim, the victim's clothing or possessions, or the location where the victim was known to have been.

✔ Victim's DNA and DNA on the suspect's body, clothing, or other possessions, or a location linked to the suspect.

Matching the evidence to the bad guy

In Figure 18-4, you can see that exact matches of DNA fingerprints stand out like a sore thumb. But when you find a match between a suspect and your evidence, how do you know that no one else shares the same DNA fingerprint as that particular suspect?

With DNA fingerprints, you can't know for sure that a suspect is the culprit you're looking for, but you can calculate the odds of another person having the same pattern. Because this book isn't *Statistical Genetics For Dummies,* I'll skip the details of exactly how to calculate these odds. Instead, I'll just tell you that, in Figure 18-4, the odds of another person having the same pattern as Suspect Two is 1 in 45 for locus one, 1 in 70 for locus two, and 1 in 50 for locus three. To calculate the total odds of a match, you multiply the three probabilities: $\frac{1}{45} \times \frac{1}{70} \times \frac{1}{50} = \frac{1}{157,500}$. Based on your test using just three loci, the probability of another person having the same DNA pattern as Suspect Two is 1 in 157,500.

When all 13 CODIS loci are used, the odds of finding two unrelated persons with the same DNA fingerprints are 1 in 53,000,000,000,000,000,000 (that's 53 quintillion for those of you scoring at home). To put this figure in perspective, consider that the planet has only 6 billion people. To say the least, your life-time odds of getting hit by lightning (1 in 3,000) are a lot better than this!

Of course, real life is a lot more complicated than the example in Figure 18-4. Biological evidence samples are often mixed and contain more than one person's DNA. Because humans are *diploid* (having chromosomes in pairs; see Chapter 2), mixed samples (called *admixtures* for you *CSI* fans) are easy to spot — they have three or more alleles in a single locus. By comparing samples, forensic geneticists can parse out whose DNA is whose and even determine how much DNA in a sample was contributed by each person.

But what if the evidence and the suspect's DNA don't match? The good news is that an innocent person is off the hook. The bad news is that the guilty party is still roaming free. At this point, investigators turn back to the CODIS system because it was designed not only to standardize which loci get used in DNA fingerprinting but also to provide a library of DNA fingerprints to help identify criminals and solve crimes. The FBI established the DNA finger-print database in 1998 based on the fact that repeat offenders commit most crimes. When a person is convicted of a crime (laws vary on which convic-tions require DNA sampling; see the sidebar "To find a criminal using DNA"), his or her DNA is sampled — often by using a cotton swab to collect a few skin cells from the inside of the mouth. As of early 2010, the CODIS database had provided over 101,000 matches and assisted in over 100,000 investiga-tions. If no match is found in CODIS, the evidence is added to the database. If the perpetrator is ever found, then a match can be made to other crimes he or she may have committed.

To find a criminal using DNA

The FBI's CODIS system works because it contains hundreds of thousands of cataloged samples for comparison. All 50 U.S. states require DNA samples to be collected from persons convicted of sex offenses and murder. Laws vary from state to state on which other convictions require DNA sampling, but so far, CODIS has cataloged over 300,000 offender samples, with at least that many more awaiting analysis. But what if no sample has been collected from the guilty party? What happens then?

Some law enforcement agencies have conducted mass collection efforts to obtain DNA samples for comparison. The most famous of these collection efforts occurred in Great Britain in the mid-1980s. After two teenaged girls were murdered, every male in the entire neighborhood around the crime scene was asked to donate a sample for comparison. In all, nearly 4,000 men complied with the request to donate their DNA. The actual murderer was captured after he bragged about how he had gotten someone else to volunteer a sample for him.

DNA evidence is also sometimes used to extend the statute of limitations on crimes when no arrest has been made. (The *statute of limitations* is the amount of time prosecutors have to bring charges against a suspect.) Crimes involving murder have no statute of limitations, but most states have a statute of limitations on other crimes such as rape. To allow prosecution of such crimes, DNA evidence can be used to file an arrest warrant or make an indictment against "John Doe" — the unknown person possessing the DNA fingerprint of the perpetrator. The arrest warrant extends the statute of limitations indefinitely until a suspect is captured.

Taking a second look at guilty verdicts

Not all persons convicted of crimes are guilty. One study estimates that roughly 7,500 persons are wrongfully convicted each year in the U.S. alone. The reasons behind wrongful conviction are varied, but the fact remains that innocent persons shouldn't be jailed for crimes they didn't commit.

In 1992, Barry Scheck and Peter Neufeld founded the Innocence Project in an effort to exonerate innocent men and women. The project relies on DNA evidence, and services are free of charge to all who qualify.

Walter D. Smith is one of the 249 persons in the U.S. exonerated by DNA evidence as of January 2010. In 1985, Smith was wrongfully accused of raping three women. Despite his claims of innocence, eyewitness testimony brought about a conviction, and Smith received a sentence of 78 to 190 years in prison. During his 11 years of incarceration, Smith earned a degree in business and conquered drug addiction. In 1996, the Innocence Project conducted DNA testing that ultimately proved his innocence and set him free.

It's unclear how many criminal cases have been subjected to postconviction DNA testing, and the success rates for such cases are unreported. Surprisingly, many states have opposed postconviction testing, but laws are being passed to allow or require such testing when circumstances warrant it.

It's All Relative: Finding Family

Family relationships are important in forensic genetics when it comes to paternity for court cases or determining the identities of persons killed in mass disasters. Individuals who are related to one another have copies of their DNA in common because each parent passes on half of his or her chromosomes to each offspring (see Chapter 5). Within a family tree, the amount of genetic relatedness, or *kinship,* among individuals is very predictable. Assuming that Mom and Dad are unrelated to each other, full siblings have roughly half their DNA in common because they each inherit all their DNA from the same parents.

Paternity testing

Surprisingly (or maybe not, depending on how many daytime talk shows you watch), roughly 15 percent of children are fathered by someone other than the father listed on the birth certificate. Therefore, tests to determine what male fathered what child are of considerable interest. Paternity testing is used in divorce and custody cases, determination of rightful inheritance, and a variety of other legal and social situations.

Paternity testing using STR techniques has become very common and relatively (pardon the pun) inexpensive. The methods are exactly the same as those used in evidence testing (see the earlier section "Outlining the powerful PCR process"). The only difference is the way the matches are interpreted. Because the STR alleles are on chromosomes (see the earlier section "Rooting through Your Junk DNA to Find Your Identity"), a mother contributes half the STR alleles possessed by a child, and the father contributes the other half. Figure 18-5 shows what these contributions may look like in a DNA fingerprint. (M is the mother, C is the child, and F1 and F2 are the possible fathers.) Alleles are depicted here as peaks, and arrows indicate the maternal alleles. Assuming that mother and father are unrelated, half the child's alleles came from F2, indicating that F2 is likely the father.

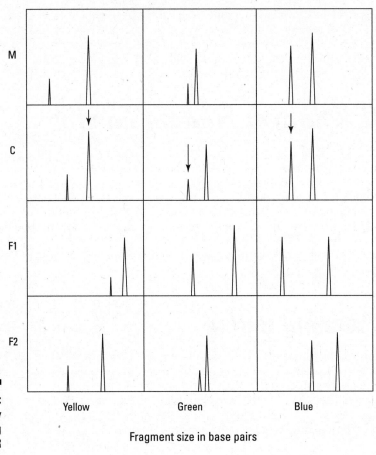

Figure 18-5:
Paternity
testing
using STR
loci.

Two values are often reported in paternity tests conducted with DNA fingerprinting:

- **Paternity index:** A value that indicates the weight of the evidence. The higher the paternity index value, the more likely it is that the alleged father is the actual genetic father. The paternity index is a more accurate estimate than probability of paternity.

- **Probability of paternity:** The probability that a particular person could have contributed the same pattern shown by the DNA fingerprint. The odds calculations for probability of paternity are more complicated than multiplying simple probabilities (see "Matching the evidence to the bad guy") because an individual who's heterozygous at a particular locus has an equal probability of contributing either allele. The probability of a particular male being the father also depends on how often the various alleles at a locus show up in the population at large (which is also true for the estimates of odds shown in "Matching the evidence to the bad guy"; see Chapter 17 for the lowdown on how population genetics works).

The results of paternity tests are often expressed in terms of "proof" of paternity or lack thereof. Unfortunately, this terminology is inaccurate. Genetic paternity testing doesn't *prove* anything. It only indicates a high likelihood that a given interpretation of the data is correct.

Thomas Jefferson's son

Male children receive their one and only Y chromosome from their fathers (see Chapter 5). Thus, paternity of male children can be resolved by using DNA markers on the Y chromosome. The discovery of this testing option led to the unusual resolution of a long-term mystery involving the third U.S. president, Thomas Jefferson.

In 1802, Jefferson was accused of fathering a son by one of his slaves, Sally Hemings. Jefferson's only acknowledged offspring to survive into adulthood were daughters, but Jefferson's paternal uncle has surviving male relatives who are descended in an unbroken male line. Thus, the Y chromosome DNA from these Jefferson family members was expected to be essentially identical to the Y chromosome DNA that Jefferson inherited from his paternal grandfather — DNA he would have contributed to a son. Five men known to have descended from Jefferson's uncle agreed to contribute DNA samples for comparison with the only

remaining male descendant of Sally Hemings's youngest son. In all, 19 samples were examined. These samples included descendants of other potential fathers along with unrelated persons for comparison. A total of 19 markers found only on the Y chromosome were used. (None of the CODIS markers is on the Y chromosome; they'd be useless for females if they were.) The Jefferson and Hemings descendants matched at all 19 markers. Since the publication of the genetic analysis, historical records have been examined to provide additional evidence that Jefferson fathered Sally Hemings's son, Eston. For example, Jefferson was the only male of his family present at the time Eston was conceived. Interestingly, examination of the historical records seems to indicate that Jefferson is likely the father of all of Sally Hemings's six children; however, this conclusion remains controversial.

Relatedness testing

Paternity analysis isn't the only time that DNA fingerprinting is used to determine family relationships. Historical investigations (like the Jefferson-Hemings case I explain in the sidebar "Thomas Jefferson's son") may also use patterns inherited within the DNA to show how closely related people are and to identify remains. Mass fatality incidents such as plane crashes and the World Trade Center disaster of September 11, 2001, rely on DNA technologies to identify deceased persons. Several methods are used under such circumstances, including STR DNA fingerprinting, mitochondrial DNA analysis (see Chapter 6 for the details on mitochondrial DNA), and Y-chromosome analysis (similar to the method I describe in the sidebar "Thomas Jefferson's son").

Several conditions complicate DNA identification of victims of mass fatality incidents. Bodies are often badly mutilated and fragmented, and decomposition damages what DNA remains in the tissues. Furthermore, reference samples of the deceased person's DNA often don't exist, making it necessary to make inferences from persons closely related to the deceased.

Reconstructing individual genotypes

Much of what forensic geneticists know about identifying victims of mass fatalities comes from airplane crashes. In 1998, Swissair Flight 111 crashed into the Atlantic Ocean just off the coast of Halifax, Nova Scotia, Canada. This disaster sparked an unusually comprehensive DNA typing effort that now serves as the model for forensic scientists the world over dealing with similar cases.

In all, 1,200 samples from 229 persons were recovered from Swissair Flight 111. Only one body could be identified by appearance alone, so investigators obtained 397 reference samples either from personal items belonging to victims (like toothbrushes) or from family members. Because most reference samples from the victims themselves were lost in the crash, 93 percent of identifications depended on samples from parents, children, and siblings of the deceased. The number of alleles shared by family members is fairly predictable, allowing investigators to conduct parentage analysis based on the expected rate of matching alleles. In the Swissair case, 43 family groups (including 6 families of both parents and some or all of their children) were among the victims, so the analyses were complicated by kinship among the victims.

The initial DNA fingerprinting of remains revealed 228 unique genotypes (including one pair of twins). The 13 CODIS loci were tested using PCR (the methods were identical to those I describe in the earlier section "Employing DNA to Catch Criminals (And Free the Innocent)"). All the data from DNA fingerprinting was entered into a computer program specifically designed to compare large numbers of DNA fingerprints. The program searched for several kinds of matches:

- ✔ A perfect match between a victim and a reference sample from a personal item
- ✔ Matches between victims that would identify family groups (parents and children, and siblings)
- ✔ Matches between samples from living family members

The computer then generated reports for all matches within given samples. Two investigators independently reviewed every report and only declared identifications when the probability of a correct identification was greater than a million to one. Altogether, over 180,000 comparisons were made to determine the identities of the 229 victims.

Forty-seven persons were identified based on matches with personal items. The remaining 182 persons were identified by comparing victims' genotypes with those of living family members. The power of PCR, combined with many loci and computer software, led to rapid comparisons and the positive identification of all the victims.

Bringing closure in times of tragedy

On September 11, 2001, two jetliners crashed into the twin towers of the World Trade Center in New York City. The enormous fires resulting from the crashes caused both buildings to collapse. Roughly 2,700 persons died in the disaster. Over 20,000 body parts were recovered from the rubble; therefore, the task of forensic geneticists was two-fold: determine the identity of each deceased person and collect the remains of particular individuals for interment. Unlike Swissair Flight 111, few victims of the WTC tragedy were related to each other. However, other issues complicated the task of identifying the victims. Many bodies were subjected to extreme heat, and others were recovered weeks after the disaster, as rubble was removed. Thus, many victim samples had very little remaining DNA for analysis.

DNA reference samples from missing persons were collected from personal effects such as toothbrushes, razors, and hairbrushes. Skin cells clinging to toothbrushes accounted for almost 80 percent of the reference samples obtained for comparison. These samples were DNA fingerprinted using PCR with the standard 13 CODIS loci I describe in "Employing DNA to Catch Criminals (And Free the Innocent)" earlier in this chapter. By July 2002, roughly 300 identifications were made using these direct reference samples. An additional 200 identifications were made by comparing victim samples to samples from living relatives using the methods I describe for the Swissair crash (see "Reconstructing individual genotypes").

By July 2004, a total of 1,500 victims had been positively identified, but subsequent progress was slow. The remaining samples were so damaged that the DNA was in very short pieces, too short to support STR analysis. Two avenues for additional identifications remained:

- ✔ **Mitochondrial DNA (mtDNA),** which is useful for two reasons:

  - It's multicopy DNA, meaning that each cell has many mitochondria, and each mitochondrion has its own molecule of mtDNA.

  - It's circular, making it somewhat more resistant to decomposition because the nucleases that destroy DNA often start at the end of the molecule (see "Collecting biological evidence" earlier), and a circle has no end, so to speak.

  mtDNA is inherited directly from mother to child; therefore, only maternal relatives can provide matching DNA. Unlike STR markers, mtDNA is usually analyzed by comparing the sequences of nucleotides from various samples (see Chapter 11 to find out how DNA sequences are generated and analyzed). Because sequence comparison is more complicated than STR marker comparison, the analyses take longer to perform but provide very accurate matches.

- ✔ **Single nucleotide polymorphism (SNP analysis)** (pronounced *snip*), which relies on the fact that DNA tolerates some kinds of mutation without harming the organism (see Chapter 13 for more about mutation). SNPs occur when one base replaces another in what's called a *point mutation.* Generally, T replaces A and G replaces C or vice versa (see Chapter 6 for more about the bases that make up DNA). These tiny changes occur often (some estimates are as high as about one in every 100 bases), and when many SNPs are compared, the changes can create a unique DNA profile similar to a DNA fingerprint.

  The downside to SNP analysis is that the point mutations don't create obvious size differences that traditional DNA fingerprinting can detect. Therefore, sequencing or gene chips (see Chapter 23 for more on gene chips) must be used to detect the SNP profile of various individuals. Because SNP analysis can be conducted on very small fragments of DNA, it allowed investigators to make more identifications than were possible otherwise. Even so, many persons were not identified, and identification efforts were halted in February 2005.

Chapter 19

Genetic Makeovers: Fitting New Genes into Plants and Animals

In This Chapter

▶ Tracing the development of genetically modified organisms

▶ Understanding how transgenics works

▶ Weighing the pros and cons of genetic modification

*O*ne of the most controversial applications of genetics technology (besides cloning, which I cover in Chapter 20) is the mechanical transfer of genes from one organism to another. This process is popularly known as *genetic modification* (GM). More properly called *transgenics,* transferring genes simplifies the production of some medications, creates herbicide-resistant plants, and has even been used to create glow-in-the-dark pets (I'm not kidding — check out the sidebar "Transgenic pets: Not all fun and games" for the details). In this chapter, you discover how scientists move DNA around to endow plants, animals, bacteria, and insects with new combinations of genes and traits.

Genetically Modified Organisms Are Everywhere

News items about genetically modified this, that, and the other crop up practically every day, and most of this news seems to revolve around protests, bans, and lawsuits. Despite all the brouhaha, genetically modified "stuff" is neither rare nor wholly dangerous. In fact, most processed foods that you eat are likely to contain one or more transgenic ingredients.

If that revelation worries you, go to your local health food store and peer at the labels on organic (and some nonorganic) foods. (*Organic* foods are generally defined as those produced without chemicals such as insecticides, herbicides, or artificial ingredients.) You'll see proclamations of "No GMO," which is meant to reassure you that no *transgenes* — genes that have been artificially introduced using recombinant DNA methods (described in Chapter 16) — were present in the plants or animals used to make the product in question.

In truth, you can't avoid genetically modified organisms in your everyday life. Genetic modification by humans, via artificial selection and, on occasion, induced mutation, created every single domesticated plant and animal species on earth. Furthermore, the ability to move genes from one species to another isn't new — viruses and bacteria do it all the time. It's a bit of a mystery as to why transgenesis is less acceptable than induced mutagenesis and artificial selection, but no matter what you call it, it's all genetic modification.

The acronyms GM (genetically modified) and GMO (genetically modified organism) are used all the time, but not in this chapter. Instead, I refer specifically to *transgenic organisms* because humans have been genetically modifying organisms in a variety of ways for a long time.

Making modifications down on the farm

Humans started domesticating plants and animals many centuries ago (take a look at the sidebar "Amazing maize" for how corn made the transition from grass to gracing your table). Historically, farmers preferentially grew certain types of plants to increase the frequency of desirable traits, such as sweeter grapes and more kernels per stalk of wheat. Many, if not all, of the cereal grains humans depend on, such as wheat, rice, and barley, are the result of selective hybridization events that created polyploids (multiple chromosome sets; see Chapter 15). When plants become polyploid, their fruits get substantially larger. Fruits from polyploids are more commercially valuable. (They're also better tasting. Try a wild strawberry if you're not convinced.)

When it comes to animals, humans purposefully inbreed various animals to increase the prevalence of traits such as high milk production in cows or retrieving ability (make that obsession) in certain breeds of dogs. (Inbreeding can also cause substantial problems; see Chapters 13 and 17 for details.)

Amazing maize

Plants depend on a variety of helpers to spread their seeds around: The wind, birds, animals, and waterways all carry seeds from one place to another. Most plants get along just fine without humans. Not so with corn. Corn depends *entirely* on humans to spread its seeds; archeological evidence confirms that corn has traveled only where humans have taken it. What's striking about this story is that modern geneticists have pinpointed the mutations that humans took advantage of to create one of the world's most widely used crops.

Primitive corn (called *maize*) put in its first appearance around 9,000 years ago. The predecessor of maize is a grass called *teosinte*. You need a good imagination to see an ear of corn when you look at the seed heads of teosinte; there's only a vague resemblance, and unlike corn, teosinte is only barely edible — it has a few rock hard kernels per stalk. Yet corn and teosinte (going by the scientific name of *Zea mays*) are the same species.

The five mutations that turned teosinte into maize popped up naturally and changed several things about teosinte to make it a more palatable food source:

✔ One gene controls where cobs appear on the plant stalk: Maize has its cobs along the entire stem instead of on long branches like teosinte.

✔ Three genes control sugar and starch storage in the kernels: Maize is easier to digest and better tasting than teosinte.

✔ One gene controls the size and position of kernels on the cob: Unlike teosinte, maize has an appearance normally associated with modern corn.

Humans apparently used teosinte for food before it acquired its mutational makeover, so it's likely that people caught on quickly to the change that developed. The mutations of the aforementioned five genes were cemented into the genome by selective harvest and planting of the new variety. People grew the mutated plants on purpose, and the only reason corn is so common now is because humans made it that way. The first true maize crops were planted in Mexico 6,250 years ago, and, as a popular addition to the diets of people in the area, its cultivation spread rapidly. Archeological sites in the United States bear evidence of maize cultivation as early as 3,200 years ago. By the time Europeans arrived, most native peoples in the New World grew maize to supplement their diets.

Relying on radiation and chemicals

In addition to domestication and selective breeding, humans have taken another path to genetically modify organisms. For over 70 years, new plant breeds have been created by purposefully induced, albeit random, mutations. In essence, plants are exposed to radiation (such as X-rays, gamma rays, and neutrons) and chemicals to produce mutant alleles aimed at producing desired traits (see Chapter 13 for how radiation damages DNA to cause mutation). Plants that commonly receive radiation and chemical treatment include

- ✔ **Food crops:** Fruits, vegetables, and grains are mutated to produce disease resistance and size and flavor variations, as well as to change the timing of fruiting. Over 2,000 different types of plants are genetically modified in this fashion. Believe it or not, you eat these varieties all the time. Ever had Rio Red grapefruit? If so, you enjoyed a mutated plant variety that acquired its deep red color from a neutron-induced mutation.

- ✔ **Ornamentals:** Many of the unusual ornamental plants you enjoy are the result of induced mutation. Roses, tulips, and chrysanthemums are all zapped to produce new flower colors.

Introducing unintentional modifications

Humans mutate plants on purpose, but we also constantly make unintentional genetic modifications on natural populations, such as mosquitoes and bacteria:

- ✔ **Mosquitoes:** Overzealous pesticide use has made most mosquito populations DDT-resistant.

- ✔ **Bacteria:** Many common antibiotics are rapidly being rendered ineffective because susceptible bacteria are wiped out, leaving only resistant strains.

These changes in bacteria and mosquito populations are due to evolutionary change; essentially, humans set up selective breeding by changing the environment. Another unintentional modification occurs when transgenes escape from controlled crops to wild plants — which they're likely to do with great frequency and efficiency. The wild plants are then genetically modified. These new, unintentional recipients of biotechnology are no less genetically modified than the crop plants (see the "Escaped transgenes" section later in the chapter).

Old Genes in New Places

If genetic modification is so ubiquitous, what's the problem with transgenic organisms? After all, humans have been at this whole genetic modification thing for centuries, right? Not exactly. Historically, humans have modified organisms by controlling matings between animals and plants with preexisting genetic compatibility.

Transgenics are often endowed with genes from very different species. (The bacterial gene that's been popped into corn to make it resistant to attack by plant-eating insects is a good example.) Therefore, transgenic organisms wind up with genes that never could have moved from one organism to another without considerable help (or massive luck; see the "Traveling genes" sidebar for more about natural gene transfer events).

After these "foreign" genes get into an organism, they don't necessarily stay put. One of the biggest issues with transgenic plants, for example, is uncontrolled gene transfer to other, unintended species. Another controversial aspect of transgenic organisms has to do with gene expression; many people worry that transgenes will be expressed in agricultural products in unwanted or unexpected ways, making foods toxic or carcinogenic.

To understand the promises and pitfalls of transgenics, you first need to know how transgenes are transferred and why. *Recombinant DNA technology* is the set of methods used for all transgenic applications. In Chapter 16, I cover the process used to find genes, snip them out of their original locations, and pop them into new locations (like the virus vectors used in gene therapy). The set of techniques used specifically to create transgenic organisms often goes by the title *genetic engineering*. Genetic engineering refers to the directed manipulation of genes to alter phenotype in a particular way. Thus, genetic engineering is also used in gene therapy to bring in healthy genes to counteract the effects of mutations.

Traveling genes

Movement of genes from one organism to another usually occurs through mitosis or meiosis, the normal mechanisms of inheritance. With *horizontal gene transfer,* genes can move from one species to another without mating or cell division. Bacteria and viruses accomplish this task with ease; they can slip their genes into the genomes of their hosts to alter the functions of host genes or supply the hosts with new, sometimes unwanted ones. This movement of genes isn't merely scientific fiction or a rare event, either. The appearance of antibiotic-resistant genes in various species of bacteria is due to horizontal gene transfers. Horizontal transfer also occurs in multicellular organisms (various species of fruit flies have shared their genes this way). One group of researchers has even shown that horizontal gene transfer may occur as a result of *eating* DNA. Yes, you read that correctly. The scientists fed mice a mixture that included DNA sequences not found anywhere in the mouse genome. The scientists found the experimentally introduced DNA circulating in the bloodstreams of their mice, strongly suggesting that horizontal transfer had actually occurred. Indeed, your own genome may owe some of its size and genetic complexity to genes acquired from bacteria. The possibility of genes turning up in unexpected places is real.

Transgenic Plants Grow Controversy

Plants are really different from animals, but not in the way you may think. Plant cells are *totipotent,* meaning that practically any plant cell can eventually give rise to every sort of plant tissue: roots, leaves, and seeds. When animal cells differentiate during embryo development, they lose their totipotency forever (but the DNA in every cell retains the potential to be totipotent; see Chapter 20). For genetic engineers, the totipotency of plant cells reveals vast possibilities for genetic manipulation.

Much of the transgenic revolution in plants has focused on moving genes to plants from bacteria, other plants, and even animals, to achieve various ends, including nutritionally enhancing certain foods, such as rice. The strongest efforts are directed at altering crops to resist either herbicides used against unwanted competitor plants or the attack of plant-eating insects.

Following the transgenesis process in plants

In general, developing transgenic plants for commercial uses involves three major steps:

1. Find (or alter) the gene that controls desired traits such as herbicide resistance.

2. Slip the transgene into an appropriate delivery vehicle (a *vector*).

3. Create fully transgenic plants that pass on the new gene along with their seeds.

Pinpointing the right gene

The process of finding and mapping genes is pretty similar from one organism to another (see Chapter 16 for some of the details). After scientists identify the gene they want to transfer, they must alter the gene so that it works properly outside the original organism. All genes must have *promoter sequences,* the genetic landmarks that identify the start of a gene, to allow transcription to occur (for the scoop on transcription, flip to Chapter 9). When it comes to creating a transgenic plant, the promoter sequence in the original organism may not be very useful in the new plant host; as a result, a new promoter sequence is needed to make sure the gene gets turned on when and where it's wanted.

Modifying the gene to reside in its new home

To date, the promoter sequences that genetic engineers use in transgenic plants are set to be always on. Therefore, the transgene's products show up in all the tissues and cell types of the entire plant in which the transgene is inserted. The all-purpose promoter often used for transgenes in plants comes from a pathogen called *cauliflower mosaic virus* (CaMV). CaMV seems to work well just about everywhere it's used and is a reliable "on switch" for the transgenes with which it's paired. When more precise regulation is needed, genetic engineers can use promoters that respond to conditions in the environment (see Chapter 11 for more about how cues in the environment can control genes).

In addition to the promoter, genetic engineers must also find a good companion gene — called a *marker gene* — to accompany the transgene. The marker gene provides a strong and reliable signal indicating that the whole unit (marker and transgene) is in place and working. Common markers include genes that convey resistance to antibiotics. With these kinds of markers, geneticists grow transgenic plant cells in a medium that contains the antibiotic. Only the plants that have resistance (conveyed by the marker gene) survive, providing a quick and easy way to tell which cells have the transgene (alive) and which don't (dead).

Getting new genes into the plant

To put new genes into plants, genetic engineers can either

- ✔ **Use a vector system from a common soil bacterium called *Agrobacterium*.** *Agrobacterium* is a plant pathogen that causes *galls* — big, ugly, tumor-like growths — to form on infected plants. In Figure 19-1, you can see what a gall looks like. Gall formation results from integration of bacterial genes directly into the infected plant's chromosomes. The bacteria enters the plant from a wound such as a break in the plant's stem that allows bacteria to get past the woody defense cells that protect the plant from pathogens (just as your skin protects you). The bacterial cells move into the plant cells (scientists aren't sure exactly how they pull off this trick), and once inside, DNA from the bacteria's *plasmids* — circular DNAs that are separate from the bacterial chromosome — integrate into the host plant's DNA. The bacterial DNA pops itself in more or less randomly and then hijacks the plant cell to allow it to replicate.

  Like the geneticists using virus vectors for gene therapy (see Chapter 16), genetic engineers snip out gall-forming genes from the *Agrobacterium* plasmids and replace them with transgenes. Host plant cells are grown in the lab and infected with the *Agrobacterium.* Because these cells are totipotent, they can be used to grow an entire plant — roots, leaves, and all — and every cell contains the transgene. When the plant forms seeds, those contain the transgene, too, ensuring that the transgene is passed to the offspring.

- ✔ **Shoot plants with a *gene gun* so that microscopic particles of gold or other metals carry the transgene unit into the plant nucleus by brute force.** Gene guns are a bit less dependable than *Agrobacterium* as a method for getting transgenes into plant cells. However, some plants are resistant to *Agrobacterium,* thus making the gene gun a viable alternative. With gene guns, the idea is to coat microscopic pellets with many copies of a transgene and by brute force (provided by compressed air) shove the pellets directly into the cell nuclei. By chance, some of the transgenes are inserted into the plant chromosomes.

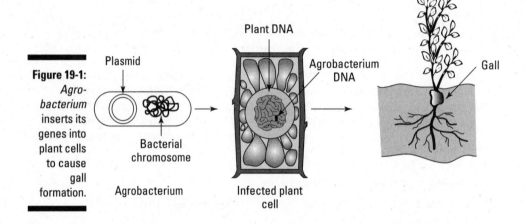

Figure 19-1:
Agro-bacterium inserts its genes into plant cells to cause gall formation.

Exploring commercial applications

Transgenic plants have made quite a splash in the world of agriculture. So far, the main applications of this technology have addressed two primary threats to crops:

- **Weeds:** The addition of herbicide-resistant genes make crop plants immune to the effects of weed-killing chemicals, allowing farmers to spread herbicides over their entire fields without worrying about killing their crops. Weeds compete with crop plants for water and nutrients, reducing yields considerably. Soybeans, cotton, and canola (a seed that produces cooking oil) are only a few of the crop plants that have been genetically altered to tolerate certain herbicides.

  A couple of different chemical companies have gotten into the transgenic plant business with the idea of producing crop plants that aren't susceptible to herbicides the company makes. The companies then market their crop plants along with their chemicals.

- **Bugs:** The addition of transgenes that confer pest-killing properties to plants effectively reduces crop losses to plant-eating bugs. Geneticists provide pest-protection traits using the genes from *Bacillus thuringiensis* (otherwise known as Bt). Organic gardeners discovered the pesticide qualities of Bt, a soil bacterium, years ago. Bt produces a protein called *Cry.* When an insect eats the soil bacteria, digestion of *Cry* releases a toxin that kills the insect shortly after its meal. Transgenic corn and cotton carry the Bt *Cry* gene; there's a potato version, too, but its cultivation was hampered because fast-food restaurants and potato chip makers refused to purchase the transgenic potatoes.

Weighing points of contention

Few genetic issues have excited the almost hysterical response met by transgenic crop plants. European efforts over the past ten years to implement the use of transgenics have been particularly contentious. Nonetheless, by 2007, 23 countries had approved transgenic crops for cultivation within their borders. Reports indicate that in some third-world countries, transgenic seeds are heavily traded on the black market as well. Opposition to transgenic plants generally falls into four basic categories, which I cover in this section.

Food safety issues

Normally, gene expression is highly regulated and tissue-specific, meaning that proteins produced in a plant's leaves, for example, don't necessarily show up in its fruits. Because of the way transgenes are inserted, however, their expression isn't under tight control (because the genes are always "on"; see "Following the transgenesis process in plants" earlier in this chapter). Opponents to transgenics worry that proteins produced by transgenes may prove toxic, making foods produced by those crop plants unsafe to eat. Researchers usually evaluate the effects of chemicals and drugs by dosing animals (usually rats and mice) with ever-increasing amounts of the chemical until they observe effects. Food products are more complicated to test, though, because test animals get not only the protein produced by the transgene but the food as well, making it hard to parse out the effects of one ingredient over another. Instead of going the megadose route with animal testing, safety evaluations of transgenic crops rely on a concept called substantial equivalence.

Substantial equivalence is a detailed comparison of transgenic crop products with their nontransgenic equivalents. This comparison involves chemical and nutritional analyses, including tests for toxic substances. If the transgenic product has some detectable difference, that trait is targeted for further evaluation. Thus, substantial equivalence is based on the assumptions that any ingredient or component of the nontransgenic product is already deemed safe and that only new differences found in the transgenic version are worth investigating. For example, in the case of transgenic potatoes, unmodified potatoes are thought to be safe, so only Bt is slated for further tests. In spite of comparison testing, researchers have had difficulty documenting any unwanted side effects from food produced with transgenic crops. Millions of persons each year consume food produced with these crops, and no ill effects have been documented thus far.

One research report published in 1999 documented the possibly hazardous nature of transgenic food. In short, the study reported evidence that rats' immune systems and organs were damaged by consuming transgenic potatoes. Upon its release, the study generated a great deal of controversy, in part because one of the authors of the study announced his findings before the paper was accepted for publication in any scientific journal. That may not sound like a big deal, but it means that experts in the field hadn't evaluated the work before it was made public. Evaluation of research results as part of the publication process is called *peer review*. Peer review is meant to prevent erroneous or bogus findings from being reported as fact. In the case of the transgenic potato uproar, the talkative author was severely castigated by the scientific community for announcing his results as valid when no evaluations other than his own had occurred. The work was eventually published, but its conclusions haven't been easy to replicate, suggesting that the result may not be valid.

Escaped transgenes

The escape of transgenes into other hosts is a widely reported fear of transgenics opponents. Canola, a common oil-seed crop, provides one good example of how quickly transgenes can get around. Herbicide-resistant canola was marketed in Canada in 1996 or so. By 1998, wild canola plants in fields where no transgenic crop had ever been grown already had not one but *two* different transgenes for herbicide resistance. This finding was quite a surprise because no commercially available transgenic canola came equipped with both transgenes. It's likely that the accidental transgenic acquired its new genes via pollination.

In 2002, several companies in the United States failed to take adequate precautions mandated by law to prevent the escape of corn transgenes via pollination or the accidental germination of untended transgenic seeds. These lapses resulted in fines — and the release of transgenes into unintended crops.

Actual transgene escape isn't widely documented yet, but containment of transgenes is virtually impossible. *Introgression,* the transfer of transgenes from one plant to another, has the potential to occur relatively frequently. Canola, sunflowers, wheat, sugar beets, alfalfa, and sorghum readily share genes with related plants. Most of these plants are wind-pollinated, meaning that mature plants easily spread their genes over very broad regions every time the breeze blows. For example, one transgenic grass used on golf courses passed its transgene for herbicide resistance on to a wild relative that was a whopping 12 miles away!

Movements of transgenes for pest and herbicide resistance may pale in the face of the newest wave of transgenic plants: pharmaceuticals. The goal of this movement is to use plants to produce proteins that were previously difficult or prohibitively costly to manufacture. Drugs to treat disease, edible vaccines, and industrial chemicals are just a few of the possibilities. As of this writing, actual field trials for some of these transgenic plants are already underway.

The consequences for transgene escape from these sorts of crops could be dire — and frankly, containment failures of other transgenic crops don't bode well for future containment prospects. And unlike Bt and herbicide-ready transgenics, the compounds produced by pharmaceuticals are truly biologically active in humans, making them truly dangerous to human health.

Developing resistance

The third major point of opposition to transgenics — the development of resistance to transgene effects — is connected to the widespread movement of transgenes. The point of developing most of these transgenic crops is to make controlling weeds or insect pests easier. Additionally, transgenic crops (particularly transgenic cotton) have the potential to significantly reduce chemical use, which is a huge environmental plus. However, when weeds or insects acquire resistance to transgene effects, the chemicals that transgenics are designed to replace are rendered obsolete.

Full-blown resistance development depends on artificial selection supplied by the herbicide or the plant itself. Resistance develops and spreads when insects that are susceptible to the pesticide transgene being used are all killed. The only insects that survive and reproduce are, you guessed it, able to tolerate the pesticide transgene. Insects produce hundreds of thousands of offspring, so it doesn't take long to replace susceptible populations with resistant ones. To counter the threat of resistance development, users of transgenic crops advocate nontransgenic refuges — places where nontransgenic crops are grown to support populations of susceptible bugs. The idea is that inheritance of the transgene resistance is diluted by the genes of susceptible bugs. So far, the implementation of refuges has seen limited success; in all likelihood, refuges may only slow the spread of resistance, not prevent it altogether.

Damaging unintended targets

The argument against transgenic plants is that nontarget organisms may suffer ill effects. For example, when Bt corn was introduced (see "Exploring commercial applications"), controversy arose surrounding the corn's toxicity to beneficial insects (that is, bugs that eat other bugs) and desirable creatures like butterflies. Indeed, Bt is toxic to some of these insects, but it's unclear how much damage these natural populations sustain from Bt plants. The biggest threat to migratory monarch butterflies is likely habitat destruction in their overwintering sites in Mexico, not Bt corn.

Assessing outcomes

Transgenic plants appear to help reduce the amount of pesticides used, but only by a small margin (between 1 and 3 percent). Since the development of transgenic plants, herbicide use has actually increased, presumably because the chemicals can be freely broadcast onto herbicide-ready crops. However, the impact of herbicide-resistant crops on *no-till,* a farming method that significantly reduces erosion and soil loss, is positive; more farmers have turned to no-till as they've adopted transgenic crops. But if weeds acquire the transgenes, this improvement will be promptly reversed. In fact, transgenic crops don't seem to have increased yields very much. Despite the relatively scant advantages and extremely strong opposition (especially in Europe), advocates of transgenic crops remain optimistic and hopeful for an agricultural revolution.

Looking at the GMO Menagerie

Transgenic critters are all over the place. Animals, insects, and bacteria have all gotten in on the fun. In this section, you take a trip to the transgenic zoo.

Transgenic animals

Mice were the organisms of choice in the development of transgenic methods. Scientists discovered that genes could be inserted into a mouse's genome during the process of fertilization. When a sperm enters an egg, there's a brief period before the two sets of DNA (maternal and paternal) fuse to become one. The two sets of DNA existing during this intermission are called *pronuclei.* Geneticists discovered that by injecting many copies of the transgene (with its promoter and sometimes with a marker gene, too; see the earlier section "Modifying the gene to reside in its new home") directly into the paternal pronucleus (see Figure 19-2), the transgene was sometimes integrated into the embryo's chromosomes. (Eggs can be injected with transgenes after the pronuclei fuse, but uptake of the transgene is somewhat less efficient.)

Not all the embryo's cells contain the transgene, however, because the uptake of the transgene takes place during cell division; sometimes, several rounds of division occur before the transgene gets scooped up. The cells that do have the transgene often have multiple copies (oddly, these end up together in a head-to-tail arrangement), and the transgenes are inserted into the mouse's chromosomes at random. The resulting, partly transgenic mouse is called a *chimera,* or a *mosaic.* Mosaicism is the expression of genes in some but not all cells of a given individual, making gene expression somewhat patchy. To get a fully transgenic animal, many chimeras are mated in the hope that homozygous transgenic offspring will be produced from one or more matings. After researchers obtain homozygotes, they isolate the transgene line so that no heterozygotes are formed by mating transgenic animals with nontransgenic animals.

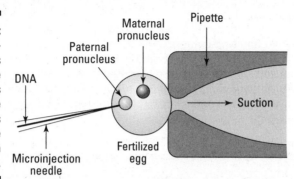

Figure 19-2: Researchers introduce transgenes into mouse embryos before fertilization occurs.

(Figure labels: Maternal pronucleus, Pipette, Paternal pronucleus, DNA, Suction, Microinjection needle, Fertilized egg)

One of the first applications of the highly successful mouse transgenesis method used growth hormone genes. Rat, human, and bovine growth hormone genes produced much larger mice than normal. The result encouraged the idea that growth hormone genes engineered into meat animals such as the following would allow faster production of larger, leaner animals:

- ✔ **Pigs and cows:** Transgenic pigs haven't fared very well; in studies, they grew faster than their nontransgenic counterparts but only when fed large amounts of protein. And female transgenic pigs turned out to be sterile. All pigs showed muscle weakness, and many developed arthritis and ulcers. Cows haven't done any better. So far, no commercially viable transgenic cows or pigs engineered for growth have been produced and, as of 2009, the U.S. authorities still haven't given the go-ahead for these animals to be part of the food supply.

- ✔ **Fish:** Unlike pigs and cows, fish do swimmingly with transgenes (see the sidebar "Transgenic pets: Not all fun and games" for one application of transgenics in fish). Transgenic salmon grow six times faster than their nontransgenic cousins and convert their food to body weight much more efficiently, meaning that less food makes a bigger fish. So far, Atlantic salmon are targeted for the growth-enhancing gene, but none are commercially available yet. Transgenic fish would be raised in pens situated in larger bodies of water, making escape of transgenics into wild populations a certainty. Plus, natural salmon populations are severely depleted due to overharvest. Farmed fish tend to be highly aggressive and are feared to out-compete their wild relatives. Thus, farm-raised salmon pose a threat to natural populations of both their own and other fish species as well. Here, too, U.S. regulators haven't given the go-ahead, and transgenic salmon aren't approved for human consumption.

Primates have also been targeted for transgenesis as a way to study human disorders including aging, neurological diseases, and immune disorders. The first transgenic monkey was born in 2000. This rhesus monkey was endowed with a simple marker gene because the purpose of the study was simply to determine whether transgenesis in monkeys was possible. The marker gene used was the one that produces green fluorescence in jellyfish. This gene has been successfully inserted into plants, frogs, and mice, but those recipients rarely glow green. The monkey recipient is no exception: Her chromosomes bear the gene, but no functional fluorescent protein is produced — yet. Twin monkeys produced in the same project died before birth, but both had fluorescent fingernails and hair follicles. Because some transgenic animals display delayed onset of the gene function, the surviving monkey may yet glow. Even with this modest success, monkeys' reproductive cycles aren't easy to manipulate, so progress with transgenic primates is slow.

CASE STUDY

Transgenic pets: Not all fun and games

Ever have one of those groovy posters that glows under a black light? Well, move that black light over to the aquarium — there's a new fish in town. Originally derived from zebrafish, a tiny, black-and-white–striped native of India's Ganges River, these glowing versions bear a gene that makes them fluorescent. The little, red, glow-in-the-dark wonders (one report calls them "Frankenfish") are the first commercially available transgenic pets.

Zebrafish are tried-and-true laboratory veterans — they even have their own scientific journal! Developmental biologists love zebrafish because their transparent eggs make it simple to observe development. Geneticists use zebrafish to study the functions of all sorts of genes, many of which have direct counterparts in other organisms, including humans. And genetic engineers have taken advantage of these easy-to-keep fish, too; scientists in Singapore saw the potential to use zebrafish as little pollution indicators. The Singapore geneticists use a gene from jellyfish to make their zebrafish glow in the dark. The action of the fluorescent gene is set up to respond to cues in the environment (like hormones, toxins, or temperature; see Chapter 11 for how environmental cues turn on your genes). The transgenic zebrafish then provide a quick and easy to read signal: If they glow, a pollutant is present.

Of course, glowing fish are so unique that some enterprising soul couldn't let lab scientists have all the fun. Thus, these made-over zebrafish have hit pet stores. Many scientists don't see the humor in making transgenic fish available to the public, however. The state of California has banned their sale outright, and at least one major pet store chain refuses to sell them. The main objection so far seems to be an ethical one — opponents object to genetic engineering used for "trivial" purposes (see Chapter 21 for details about ethics and genetics). The U.S. Food and Drug Administration (FDA), however, has deemed glowing zebrafish safe (they're nontoxic, and no, you won't glow if you eat one).

A more serious and biologically relevant argument against glowing exotic fish may be the threat of invasive species. Invasive species present all kinds of nasty problems for the environment. For example, the reason you don't enjoy home-grown chestnuts in the United States anymore is because an introduced plant disease literally wiped out every single tree. Introduced insects, plants, and animals represent an enormous and expensive threat to agriculture worldwide. Regular zebrafish already live in Florida's warm waters, along with a dizzying number of other nonnative fish that collectively threaten to destroy the native fish community entirely. Glowing fish may be only the beginning, by the way. Reports suggest that glow-in-the-dark lawn grasses and grasses in unusual colors are in the works. And remarkable glowing colors are only one possibility. One company has announced plans to make hypoallergenic cats! (But don't hold your breath; animals don't respond well to the random insertion of genes into their chromosomes, so the production of sneeze-free kitties is a distant dream.)

Trifling with transgenic insects

A number of uses for transgenic insects appear to be on the horizon. Malaria and other mosquito-borne diseases are a major health problem worldwide, but the use of pesticides to combat mosquito populations is problematic because resistant populations rapidly replace susceptible ones. And in fact, the problem isn't really the mosquitoes themselves (despite what you may think when you're being buzzed and bitten). The problem is the parasites and viruses the mosquitoes carry and transmit through their bites. In response to these problems, researchers are developing transgenic mosquitoes unable to carry parasites or viruses, rendering their bites itchy but otherwise harmless. Unfortunately, it's not clear how or if transgenic mosquitoes could replace populations of bugs that carry diseases.

Other attempts at biological control of insects have met with limited success. They usually involve the release of millions of sterile bugs that attract the mating attentions of fertile ones. The matings result in infertile eggs, reducing the reproduction of the target insect population. Part of the downside of this environmentally friendly approach to pest control is that sterility is induced using radiation, and irradiated insects lack the vigor needed to aggressively pursue sex. Transgenic infertility may solve the problem. The general process is the same, but the transgenically infertile insects still have the energy needed to pursue mates, resulting in a more effective pest control strategy. This is an especially appealing idea when used to combat invasive species that can sweep through crops with economically devastating results.

The whole transgenic pesticide-resistance affair may be used to enhance natural control of pest populations by using insects that make a living eating other bugs. The idea is to create beneficial insects that bear the transgene that confers pesticide resistance. The farmers can then put out pesticides to kill susceptible insects and release beneficial bug predators to do the rest. Such a strategy may reduce pesticide use dramatically and eliminate the need for transgenic, insect-resistant crops.

Another transgenic insect project in the works uses silkworms that are equipped with a gene used to make human skin protein. The intention is to mass produce the protein for use in human skin grafts needed after burns and to aid with wound healing.

Fiddling with transgenic bacteria

Bacteria are extremely amenable to transgenesis. Unlike other transgenic organisms, genes can be inserted into bacteria with great precision, making expression far easier to control. As a result, many products can be produced using bacteria, which can be grown under highly controlled conditions, essentially eliminating the danger of transgene escape. (The techniques used to slip genes into bacteria chromosomes are identical to those used in gene therapy, which I describe in Chapter 16.)

Many important drugs are produced by recombinant bacteria, such as insulin for treatment of diabetes, clotting factors for the treatment of hemophilia, and human growth hormone for the treatment of some forms of dwarfism. These sorts of medical advances can have important side benefits as well:

✔ Transgenic bacteria can produce much greater volumes of proteins than traditional methods.

✔ Transgenic bacteria are safer than animal substitutes, such as pig insulin, which are slightly different from the human version and therefore may cause allergic reactions.

✔ Transgenic bacteria are much less controversial than other organisms and thus are well received for the production of medications.

Transgenic bacteria are also used for applications down on the farm. *Bovine somatotropin,* better known as bovine growth hormone, increases milk production in cows. Transgenic bacteria are used to produce large quantities of the hormone (called rbGH for recombinant bovine growth hormone), which is injected into dairy cows to boost milk production. Despite outcries to the contrary, studies show that rbGH isn't active in humans, meaning that humans don't respond to bovine growth hormone even when it's injected in their bodies. Furthermore, milk produced by cows injected with rbGH is chemically indistinguishable from milk produced by cows injected with the actual hormone. The advantage of rbGH is that it allows fewer cows to produce more milk — a good thing, because dairies represent a significant source of fecal pollution in rivers and streams, and fewer cows means less pollution. The downside is that cows treated with rbGH are more vulnerable to infection, requiring treatment with antibiotics and thus increasing the risk of developing antibiotic-resistant bacteria.

Recent advances in biotechnology may produce other gains in protecting the environment. For example, work is underway to take advantage of the production of biodegradable plastics using bacteria-produced chemicals called *polyhydroxyalkanoates* (PHAs). PHAs are molecules that are used like fats to store energy. They're also very similar to the plastics made from petroleum that you see all the time. Researchers have taken the gene that makes PHA and popped it into *E. coli* to produce enough PHA with which to manufacture products. And it's likely that PHAs will find their way into the marketplace as a viable alternative to traditional plastics.

Chapter 20

Cloning: You're One of a Kind

In This Chapter

▶ Defining cloning

▶ Understanding how cloning works

▶ Looking at some common clone abnormalities

▶ Sorting out the arguments for and against cloning

*1*t sounds like science fiction: Harvest your genetic information, implant that information into an egg cell, and after nine months, welcome a new baby into the world. A new baby with a difference — it's a clone.

Depending on your point of view, cloning organisms may sound like a nightmare or a dream come true. Whatever your opinion, cloning is most definitely not science fiction; decisions about experimental cloning are being made right now, every day. This chapter covers cloning: what it is, how it's done, and what its impact is from a biological point of view. You get to know the problems inherent in clones, along with the arguments for and against cloning (not just of humans — of animals and plants, too). Get ready for an interesting story, and remember, it ain't fiction!

Send in the Clones

A *clone* is simply an identical copy. When geneticists talk about cloning, they're most often talking about copying some part of the DNA (usually a gene). Geneticists clone DNA in the lab every day — the technology is simple, routine, and unremarkable. Cloning genes is a vital part of

✔ DNA sequencing (see Chapter 8)

✔ The study of gene functions (see Chapter 11)

✔ The creation of recombinant organisms (see Chapter 16)

✔ The development of gene therapy (see Chapter 16)

Another meaning of the word *cloning* is to make a copy of an entire organism as a reproductive strategy. When referring to a whole creature as opposed to DNA, a clone is an organism that's created via asexual reproduction, meaning that offspring are produced without the parent having sex first. Cloning occurs naturally all the time in bacteria, plants, insects, fish, and lizards. For example, one type of asexual reproduction is *parthenogenesis,* which occurs when a female makes eggs that develop into offspring without being fertilized by a male (for some of you female readers, I'm sure this sounds very appealing). So if reproduction by cloning is a natural, normal biological process, what's the big deal with cloning organisms using technology?

Cloning Animals: Like No Udder

Cloning animals hit the news big time in 1997 with the birth of Dolly, an unremarkable looking Finn Dorset lamb. Named after the well-endowed country singer Dolly Parton, Dolly the sheep was a clone of one of her mother's udder cells. (If you didn't grow up on a farm, *udders* are the part of the animal that produces milk — in other words, breasts. Hence the name of everyone's favorite clone.) I use the term "mother" rather loosely when it comes to Dolly; the cells came from one animal, the egg was derived from a second animal, and yet a third female was the birth mom.

Dolly's name was intended as a bit of a joke, but the fact that an animal had been cloned meant many people weren't laughing. Images of a future filled with mass-produced human beings began to fill the minds of many. Clones aren't unique individuals and, in cases like Dolly's, are produced via technology. Therefore, human rights advocates and religious leaders often object to cloning on moral or ethical grounds (see "Arguments against cloning" later in this chapter).

Despite her ordinary appearance, Dolly was unique in that no other mammal had been reproduced successfully via cloning using a somatic (body) cell (see the section "Discovering why Dolly is really something to bah about" later in the chapter). But Dolly wasn't the first organism to be cloned.

Cloning before Dolly: Working with sex cells

Experimental cloning started in the 1950s. In 1952, researchers transplanted the nucleus from a frog embryo into a frog egg. This and subsequent experiments were designed not to clone frogs but to discover the basis of *totipotent cells.* Totipotent cells are capable of becoming any sort of cell and are the basis for all multicellular organisms. Totipotency lies at the heart of developmental genetics.

For most organisms, after an egg is fertilized, the zygote begins developing by cell division, which I walk you through in Chapter 2. Division proceeds through 2, 4, 8, and 16 cells. After the zygote reaches the 16-cell stage, the cells wind up in a hollow ball arrangement called a *blastocyst.* Figure 20-1 shows the stages of development from 2 cells to blastocyst.

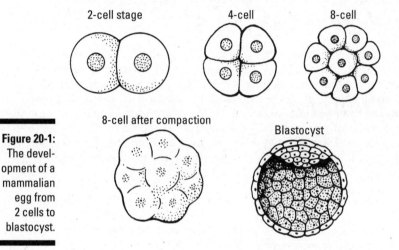

2-cell stage   4-cell   8-cell

8-cell after compaction

Blastocyst

Figure 20-1:
The development of a mammalian egg from 2 cells to blastocyst.

Zygote development in mammals is unique because the cells don't all divide at the same time or in the same order. Instead of proceeding neatly from 2 to 4 to 8, the cells often wind up in odd numbers. Mammal zygotes have a unique stage of development, called *compaction,* when the cells go from being separate little balls into a single, multicellular unit (refer to Figure 20-1). Compaction occurs after the third round of cell division. After compaction, the cells divide again (up to roughly 16), the inner cell mass forms (this will become the fetus), and fluid accumulates in the center of the ball of cells to form the blastocyst.

After a few more divisions, the cells rearrange into a three-layered ball called a *gastrula.* The innermost layer of the gastrula is *endoderm* (literally "inner skin"), the middle is *mesoderm* ("middle skin"), and the outermost is *ectoderm* ("outer skin"). Each layer is composed of a batch of cells, so from the gastrula stage onward, what cells turn into depends on which layer they start out in. In other words, the cells are no longer totipotent; they have specific functions.

So why is totipotency important? The entire body plan of an organism is coded in its DNA. Practically every cell gets a copy of the entire body plan (in the cell nucleus; see Chapter 6 for more on DNA and cells). Despite having access to the entire genome, however, eye cells produce only eye cells, not blood cells or muscle cells. All cells arise from totipotent cells but end up *nullipotent* — able to produce only cells like themselves. Totipotent cells hold the key to gene expression and what turns genes on and off (which I cover in detail in Chapter 11). Understanding what controls totipotency also has broad implications for curing diseases such as cancer (see Chapter 14), treating spinal cord injuries using totipotent stem cells (see Chapter 23), and curing inherited disorders (see Chapter 15).

Discovering why Dolly is really something to bah about

The scientific breakthrough that Dolly the sheep signifies isn't cloning. The real breakthrough is that Dolly started off as a nullipotent cell nucleus. For many years, scientists couldn't be certain that loss of totipotency didn't involve some change at the genetic level. In other words, researchers wondered if the DNA itself got altered during the process of going from totipotent to nullipotent. Dolly convincingly demonstrated that nuclear DNA is nuclear DNA regardless of what sort of cell it comes from. (See Chapter 6 for more about nuclear DNA.) Theoretically, *any* cell nucleus is capable of returning to totipotency. That may turn out to be very good news.

The promise of *therapeutic cloning* is that someday doctors will be able to harvest your cells, use your DNA to make totipotent cells, and then use those cells to cure your life-threatening disease or restore your damaged spinal cord to full working order. However, creating totipotent cells from nullipotent cells to treat injury or disease is difficult and triggers significant ethical debates (see "Fighting the Clone Wars" later in this chapter), so realizing the potential of therapeutic cloning may be a very long way off. Meanwhile, *reproductive cloning* — the process of creating offspring asexually — is already causing quite a stir. For a taste of some of the excitement, see the sidebar "Aclone in the universe?"

Aclone in the universe?

When a human clone was reportedly born in December 2002, the news wasn't wholly unexpected. A company called Clonaid made the announcement and, purportedly, the clone; the company claimed that one cloned child had been born and numerous clone babies were on the way. The question is: On their way from where? As it turns out, Clonaid was founded by a group called the Raelians. While being entertained by sexy robots on board an alien spacecraft, Rael, the group's founder, reportedly learned that all humans are descended from clones created 25,000 years ago by space aliens.

But Clonaid's claims (all of them, including the sexy robot thing) are unsubstantiated. Shortly after the initial announcement of cloning success, Clonaid was invited to submit samples for genetic testing to support its claim. Ultimately, Clonaid refused genetic tests as an infringement on the right to privacy by the cloned child's parents. There's no word on how many parents the cloned child may have (with the egg mother, womb mother, and cell donor, it could be as many as three different people!). At the time of this writing, Clonaid continues to tout human cloning services for which it reportedly charges $200,000. Among other interesting tidbits, its Web site says that the company found living cells in a body that was dead for over four months, giving a whole new meaning to the "living dead."

Creating Clones

Despite the fact that rats, mice, goats, cows, horses, pigs, and cats have all been cloned, cloning isn't easy or routine. Cloning efficiency (the number of live offspring per cloning attempt) is generally very low. Dolly, for example, was the only live offspring out of 277 tries. All sorts of other biological problems also arise from cloning, but to understand them, you first need to understand how clones are created. (I return to the subject of challenges and problems in the "Confronting Problems with Clones" section later in the chapter.)

Making twins

One simple way to make a clone is to take advantage of the natural process of twinning. Identical twins normally arise from a single fertilized egg, called a *zygote* (see the "Cloning before Dolly: Working with sex cells" section earlier in the chapter). The zygote goes through a few rounds of cell division, and then the cells separate into two groups, each going on to form one offspring.

Artificial twinning is relatively simple and was first done successfully (in sheep) in 1979. A single fertilized egg was used, meaning that the offspring was the result of sexual reproduction. Zygotes from normally fertilized (sexually produced) eggs were harvested from ewes (female sheep). The zygote was allowed to divide up to the 16-cell stage (see the "Cloning before Dolly: Working with sex cells" section earlier in the chapter). The 16 cells were then divided into two groups, which went right on dividing, and after they were implanted into the reproductive tract of the ewe, they resulted in twins. The twins were genetically identical to each other because they were produced from the same fertilized egg.

In cows, about 25 percent of artificial embryo splits result in twin births; 75 percent result in only one calf. Nonetheless, the procedure is successful enough to increase the number of calves by about 50 percent over conventional fertilizations. This sort of cloning is relatively routine in agricultural settings and has received surprisingly little attention in the debate over cloning. The fact that the clones arise from a fertilized egg may have dampened the furor somewhat.

Using a somatic cell nucleus to make a clone

Somatic cells are body cells. Typically, body cells are *nullipotent,* meaning they only make more of the same kind of cell by mitosis (see Chapter 2 for all the details on mitosis). For example, your bone cells only make more bone cells, your blood cells only make more blood cells, and so on. Most somatic cells have nuclei that contain all the information needed to make an entire organism — in the case of cloning, a clone of the cell's owner (sometimes referred to as a *donor*).

Harvesting the donor cell

The choice of cell type used for cloning isn't trivial. The cells must grow well *in vitro* (literally "in glass," as in a test tube), and those from the female reproductive tract (mammary, uterine, and ovarian cells) seem to work best. Sorry, guys, but so far, very few clones are male.

Because body cells are often in the process of dividing (mitosis), the donor's cell must be treated to stop cell division and leave the cell in the G0 stage of mitosis (see Chapter 2). In this state, the chromosomes are "relaxed," and the DNA isn't undergoing replication. After the cell is made inactive, the nucleus of the donor cell and all the chromosomes inside it are harvested. This harvest is usually accomplished by gently drawing the cell nucleus out with a needle attached to a syringe-like tool called a *pipette*. The process of removing a cell nucleus is called *enucleation,* and the resulting cell is *enucleated* (that is, lacking a nucleus).

Harvesting the egg cell

To complete the process of making a clone using the somatic cell method, another cell is needed — this time, an egg cell. Egg cells are generally the largest cells in the body. In fact, a mature mammalian egg cell is visible to the naked eye; it's about the size of a very small speck of dust, like what you might see floating in the air when a shaft of sunlight pierces an otherwise dark room.

To harvest an egg cell, the female animal (here, called the *egg mother*) is treated with a hormone to stimulate ovulation. When the egg mother produces eggs, she first makes an *oocyte,* or immature egg (see Chapter 2 for a full rundown of egg production as part of gametogenesis). At the oocyte stage, the egg has completed the first round of meiosis (meiosis I) but isn't ready to be fertilized. The oocyte is harvested, and all the chromosomes are removed (oocytes don't really have nuclei to contain chromosomes) using the same method used for the somatic cell, leaving only the cytoplasm behind (take a peek at Figure 20-2). Also remaining in the oocyte's cytoplasm are mitochondria, which each contain a copy of the egg mother's mitochondrial DNA (see Chapter 6). After the clone is formed, the egg mother's mitochondrial DNA and the donor cell's nuclear DNA may interact and have unexpected consequences (see "Confronting Problems with Clones" later in this chapter).

As it turns out, some egg cells are really versatile. Rabbit egg cells have been used to clone cats, for example. Generally, though, staying within a species works best — that is, cat egg cells work best with cat somatic cell nuclei. See the sidebar "Clone, Spot, clone!" for more about cloned kitties.

Putting it all together

With both the donor cell and the egg cell in hand, the nucleus from the donor cell is injected into the enucleated oocyte (see Figure 20-2). The donor nucleus is fused with the oocyte using a brief electrical shock. This little jump-start plays the part of fertilization: The oocyte starts dividing and begins developing into an embryo. After cell division is well established, the dividing cells are implanted into a female (the birth or gestation mother) for the remainder of the pregnancy. Dolly the sheep clone was born after 148 days of gestation, which is about 5 days longer than average for a Finn Dorset sheep.

Clone, Spot, clone!

Yes, folks, it's possible to clone your kitty or duplicate your doggie. That was the promise of a company called Genetic Savings and Clone, which offered tissue preservation services and cloning. The first cloned cat, named CC for Copy Cat, was produced by researchers at Texas A&M University in 2002 (the ultimate revenge for all those Aggie jokes, I guess). The work, funded by billionaire John Sperling, was originally meant for canine cloning. Sperling wanted to clone his own beloved dog, Missy, who died in 2002. Like most cloning efforts, success rates are low; only one in 87 attempts produce a live kitten. But as it turns out, that old phrase "copy cat" has a deeper meaning — cats are a lot easier to clone than dogs for a number of reasons. Dogs' reproductive biology isn't very amenable to the forced ovulation required for oocyte harvesting. Unfortunately, demand for cloned pets wasn't sufficient to keep Genetic Savings and Clone in business; they folded in 2006.

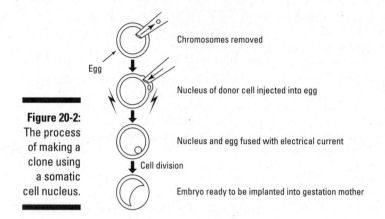

Chromosomes removed

Egg

Nucleus of donor cell injected into egg

Nucleus and egg fused with electrical current

Cell division

Embryo ready to be implanted into gestation mother

Figure 20-2:
The process
of making a
clone using
a somatic
cell nucleus.

Confronting Problems with Clones

At birth, Dolly seemed normal in every way. She grew to adulthood, was mated to a ram, and gave birth to her own lambs (a total of six over her lifetime). However, Dolly lived only 6 years; normally, Finn Dorsets live 11 or 12 years. Dolly became ill with a lung disease and was euthanized to relieve her suffering. The first hint that Dolly wasn't completely normal was arthritis. She developed painful inflammation in her joints when she was only 4 years old. Arthritis isn't unusual in sheep, but it usually only occurs in very old animals.

As it turns out, a number of abnormalities are common among clones. Clones suffer from a variety of physical ailments, including heart malformations, high blood pressure, kidney defects, impaired immunity to diseases, liver disorders, malformed body parts, diabetes, and obesity. The following sections examine the most common physical problems clones face.

Faster aging

Before somatic cells divide during mitosis, the DNA in each cell must replicate (see Chapter 7 for replication information). Each chromosome is copied except for the chromosomes' ends, called *telomeres,* which aren't fully replicated. As a result, telomeres shorten as the cell goes through repeated rounds of mitosis. Shortening of telomeres is associated with aging because it happens over time (see the sidebar "Your aging DNA" for more). Telomere shortening may mean problems for clones created through somatic nucleus transfer because, in essence, such clones start out with "aged" DNA.

Dolly the cloned sheep had abnormally short telomeres, giving rise to the worry that perhaps all clones may suffer from degenerative diseases because of premature aging. Research with other clones has provided conflicting results. Some clones, like Dolly, have shortened telomeres. Surprisingly, some clones seem to have reversed the effects of aging; specifically, their telomeres are repaired and end up longer than those of the donor. What this reversal suggests is that embryonic cells have *telomerase,* the enzyme that builds new telomeres using an RNA template during DNA replication (see Chapter 7 for more about telomerase and its role in replication). In the end, the possibility of premature aging in clones is a real one, but not all clones seem to be susceptible to it.

Your aging DNA

As you get older, your body changes: You get wrinkles, your parts start to sag, and your hair goes gray. Eventually, the chromosomes in some of your cells get so short that they can no longer function properly, and the cells die. This progressive cell death is thought to cause the unwelcome signs of aging you're familiar with. In fact, the shortening of telomeres in most animals is so predictable that it can be used to determine how old an animal is.

All your cells have the genes to make telomerase (see Chapter 7). But telomerase genes are turned on only in certain kinds of cells: germ cells (those that make eggs and sperm), bone marrow cells, skin cells, hair follicle cells, and the cells that line the intestinal walls (in other words, cells that divide a lot). Cancer cells also have telomerase activity, a fact that allows the unregulated growth of tumors that's sometimes fatal (see Chapter 14 for more on genetics and cancer).

In experiments, mice without a functioning telomerase gene age faster than normal mice.

This finding led some researchers to believe that telomerase may be used (eventually) to reverse or prevent aging in humans. Recently, however, research shows that telomere length is only part of the story. Telomeres interact with proteins that cover them and act as caps. When those protein caps are missing, the cell cycle gets disrupted and may stop altogether, causing premature cell death. Finally, stress may play a significant role in how fast telomeres shorten. A study of mothers with chronically ill children showed that signs of aging were accelerated in the moms of ill children compared to moms of the same age with healthy children. The stressed moms had shortened telomeres and, from a cellular point of view, were up to 10 years older than their actual ages. Although telomerase may someday be part of treating stress and aging, research indicates that the best bet against aging may be lowering stress levels the old-fashioned way: rest and relaxation.

Bigger offspring

Clones tend to be physically large; at birth, they have higher than average weight and larger than normal body size. Many clones, such as cows and sheep, must be delivered by cesarean section because they're too large to be born naturally. In part, the large birth size of clones is due to the fact that they stay in the womb longer than usual. Dolly the cloned sheep, for example, was born about five days after her birth mother's "due date." Offspring not born shortly after the normal due date (in humans, about two weeks late) are at great risk for stillbirth and complications, such as difficulty breathing. Clones tend to have very large *placentas* (the organ that links fetus to mother for oxygen and nutrition), which may contribute to their larger size, but the exact reason for the longer gestation periods is unclear.

The problem with oversize clones is so pervasive that it's been dubbed *large offspring syndrome,* or LOS. Many offspring, including humans, produced using *in vitro fertilization* (so-called "test-tube babies") also suffer from LOS, suggesting that it's not necessarily a problem associated with cloning. Instead, LOS seems to result from manipulation of the embryo. These manipulations cause changes in the way genes for growth are expressed (see Chapter 11 for more about gene expression).

Genomic imprinting occurs when genes are expressed based on which parent they come from. (For more on genomic imprinting, jump to the sidebar "It takes two to make a baby.") In the case of LOS, what seems to happen is that the genes derived from the most recent male ancestor tell the fetus to grow faster and bigger than normal. Normally, genomic imprinting affects fewer than 1,000 genes (out of 22,000 or so total genes in humans; see Chapter 8). How these "paternal" genes get turned on is anybody's guess, but the interaction of sperm with egg during fertilization is likely part of the answer. The end result of LOS is large offspring that often suffer from a variety of birth defects and are at risk for certain kinds of cancer. Estimates of LOS in human children born by *in vitro* fertilization are about 5 percent. (Normally, LOS occurs in less than 1 percent of children produced through natural fertilization.)

It takes two to make a baby

Needing a mom and a dad to make a baby sounds like common sense, but the wonders of genetic engineering suggest otherwise (after all, Dolly had three mothers and no father). Maternal and paternal DNA *are* required for successful reproduction — at least by mammals — because of genomic imprinting.

Genomic imprinting was first discovered in studies with mice. Researchers created mouse embryos with DNA from either female or male mice, but not both. Embryos with only paternal or maternal DNA didn't develop normally, indicating that both male and female DNA are required for successful development. In other studies, mice were engineered to have certain genes (see Chapter 19 for more on transgenic animals). The expression of the genes in offspring of the transgenic mice depended on which parents transmitted the genes. All offspring inherited the genes, but the genes were expressed only when the fathers transmitted them. Likewise, certain genes were expressed only when transmitted by the mothers. Thus, the growth and development of offspring is regulated by genes turned on simply because they come from mom or dad. Those genes then act in concert to regulate normal development of the embryo.

Developmental disasters

The percentage of cloning attempts that result in live births is extremely low. Generally, scientists must carry out hundreds of cell transfers to produce one offspring. Most clones perish immediately because they never implant into the gestation mother's uterus. Of the embryos that do implant and begin development, more than half die before birth. In many cases, the placenta is malformed, preventing the growing fetus from obtaining proper nutrition and oxygen.

In most cloning attempts, two females are involved. The egg comes from one female and gets implanted into another female for gestation. Therefore, another cause of early death may be that the gestation mother rejects the clone as foreign. In these cases, the gestation mother's immune system doesn't recognize the embryo as her own (because it's not) and secretes antibodies to destroy it. *Antibodies* are chemicals that the body produces to interact with bacteria, viruses, and foreign tissues to fight disease.

Some of the problems suffered by clones may result from the mismatch between mitochondrial and nuclear DNA. When an oocyte is harvested from a female different from the somatic cell (see the earlier section "Using a somatic cell nucleus to make a clone"), the egg contains roughly 100,000 copies of the egg mother's mitochondrial DNA. Unless the donor cell comes from the egg mother's sister, the somatic cell nucleus comes from a cell with a different mitochondrial genome. This mismatch means that the clone isn't a true clone — its DNA differs slightly from the donor. Cloned mice with mismatched mitochondrial and nuclear DNA tend to have decreased growth rates compared to cloned mice with matched mitochondrial and nuclear genomes.

The type of donor cell used also makes a difference in the clone's health. When introduced into the oocyte, the donor cell nucleus gets "reprogrammed" somehow to go from nullipotent to totipotent. Some cell nuclei seem to be better at resetting to totipotent than others. Almost all clones whose genomes don't get reprogrammed perish.

Effects of the environment

Clones are *never* truly exact copies of the donor organism, because genes interact with the environment in unique ways to form phenotypes, or physical qualities. If you've ever known a set of identical twins, you know that twins are very different from each other. *Monozygotic* (a fancy term for "identical" that literally means "one egg") twins have different fingerprints, develop at different rates, have different preferences, and die at different times. Being genetically identical doesn't mean they're truly, 100 percent identical.

The environment's role in development is perhaps best illustrated by experiments using plants. Suppose shoots from a single plant are rooted and grown at different locations. In essence, the plants are clones of the parent plant. If genetic control were perfect, we'd expect identical plants to perform in identical ways, regardless of environmental conditions. However, the plants in our experiment grow at very different rates depending on their location. In other words, identical plants perform differently under different conditions. Likewise, genetically identical mice raised under exactly the same conditions don't respond in identical ways to exactly the same doses of medications.

All organisms respond to their environment in unique and unpredictable ways. From the very beginning, animals experience unique conditions inside the womb. Hormonal exposure during pregnancy can have profound effects on developing organisms. For example, female piglets sandwiched between brothers while in the womb are more aggressive as adults than females that were situated between sisters. This is because male piglets secrete testosterone — a hormone that increases aggressive behavior.

Attempts to replicate organisms exactly are doomed to failure. Genetics doesn't control destiny, because genes aren't expressed in predictable ways. Persons carrying mutations for certain diseases don't have a 100-percent probability of developing those diseases (see Chapter 13). Likewise, clones don't express their genes in precisely the same way as the donor organism. Add the differences in mitochondrial DNA, *in utero* conditions (clones usually develop in a different womb), and time periods to the huge differences already present, and the only conclusion is that no clone will ever experience the world in precisely the same way as the donor organism did.

Fighting the Clone Wars

The arguments for and against cloning are numerous. In the sections that follow, I review some of the main points in both the pro and con corners. As you read, understand that these aren't *my* opinions and arguments; I only summarize what others have argued before me. I try to be balanced and fair, because before you can responsibly take a position on cloning, you need to know both sides of this controversial topic. And for more information on ethical considerations in genetics, see Chapter 21.

Arguments for cloning

Like every other scientific discovery, cloning can be used to do a lot of good. Cloning for medical and therapeutic purposes gives enormous hope that paralyzed persons will walk again and that people suffering from previously incurable conditions such as muscular dystrophy and diabetes will be cured. Cloning has provided scientists with some important answers about how genetics works. Prior to these discoveries, the changes that occur from embryo to adult were believed to cause permanent changes to the organism's DNA. Now we know that's not true. Because all DNA has the potential to return to totipotence, doctors have the unparalleled opportunity to correct genetic defects and provide treatment for devastating progressive diseases.

Another plus in the pro-cloning camp is that cloning may provide genetically matched organisms that will streamline research into the causes and treatments of diseases such as cancer. Because matched comparisons are scientifically more powerful, fewer animals are needed to conduct experiments. Such changes are an important advance over current research methods and will improve conditions for experimental animals.

Advancing knowledge of genetics can provide dramatic benefits not only to humans but also to the planet as a whole. Cloning may represent the last hope for some rare and endangered species. When only a few individuals remain, cloning may provide additional individuals to allow the population to survive. Given that the earth is experiencing its largest wave of species extinctions since ancient times, cloning may be a very significant advance for conservation biology.

Arguments against cloning

Although cloning represents an enormous opportunity, it's an opportunity fraught with danger. For the first time in history, humans possess the technology to create genetically modified organisms. That capability extends not just to animals and plants but to humans as well. Furthermore, the genetic diversity that gives the natural world its rich texture is endangered by a unique threat — that of creating organisms that are genetically identical.

As I discuss in Chapter 17, genetic diversity is extremely important to establishing and maintaining the health and well-being of populations of organisms. Research shows that genetically diverse populations are more resilient to environmental stress and are better at resisting disease. Thus, creating populations of genetically similar organisms exposes all organisms to greater threats of disease. Lack of genetic diversity in populations of other organisms may ultimately expose humans to threats as well. For example, genetically identical crops could all fall prey to the same disease and, consequently, seriously endanger food supplies (this isn't as farfetched as it sounds). In fact, efforts to archive genetically diverse strains of plants are already underway lest unique genetic characteristics, like disease resistance, are lost.

Furthermore, cloning is fraught with problems for which no good alternatives exist. For now, all cloning requires oocytes from female organisms. Those oocytes are obtained by first treating females with large doses of fertility drugs to stimulate ovulation. Such drugs stress the female's system enormously and increase the rate of cell turnover in her ovaries. Some studies indicate that the drugs used for stimulating ovulation expose females to increased risk of ovarian cancer. And the risk doesn't end there. When eggs are produced, they must be surgically removed under anesthesia. Regardless of the precautions, the female organism can and does experience pain. Animals can't give or withhold consent, so they're subjected to these procedures whether they like it or not.

After eggs are harvested and donor cells are fused with them, development of an embryo begins. The vast majority of cloning attempts, regardless of their ultimate purpose, result in death of the embryo. Granted, these embryos have no nerve cells and no consciousness that scientists know of, but nevertheless, living organisms are produced with little or no hope of survival.

If clones are successfully created, their quality of life may be poor. Clones suffer from a myriad of disorders for which causes are unknown. They may age prematurely and are likely at risk for disorders that are yet unrecognized consequences of the methods used in the cloning process. Like the experimental animals used for egg production, cloned animals can't withhold their consent and withdraw from study.

The most contentious issue posed by cloning technology is the production of human clones. As with animals, most cloned human embryos would have no hope of survival. Women must consent to painful and potentially dangerous procedures to produce eggs, and some women must consent to carry the developing child and risk the emotional trauma of miscarriage or stillbirth. From an emotional standpoint, children created this way would be genetically identical to some other person, whether that person is living or dead. The pressure to be like someone else would undoubtedly be enormous. Further, because of the genetic similarities to some other individual, parents may have unrealistic expectations of their cloned offspring. Do individual humans have a right to genetic uniqueness? It's a difficult question, but it's one we need to answer soon, before human cloning becomes true reality.

Chapter 21

Giving Ethical Considerations Their Due

In This Chapter

▶ Examining the dark side of genetics

▶ Pushing the envelope of informed consent

▶ Protecting genetic privacy

The field of genetics grows and changes constantly. If you follow the news, you're likely to hear about several new discoveries every week. When it comes to genetics, the amount of information is bewildering, and the possibilities are endless. If you've already read many of the chapters in this book, you have a taste of the many choices and debates created by the burgeoning technology surrounding our genes.

With such a fast-growing and far-reaching field as genetics, ethical questions and issues arise around every corner and are interconnected with the applications and procedures. Throughout this book, I highlight this interconnectedness. I cover animal welfare issues (in the context of cloning) in Chapter 20. Conservation of the environment and endangered species is a key part of the discussion of population genetics in Chapter 17. Chapter 19 touches on the potential dangers — to the environment and to humans — of genetic engineering. Genetic counseling, including some of the issues surrounding prenatal testing, is the subject of Chapter 12. And in Chapter 16 I discuss gene therapy as an experimental form of treatment.

But I couldn't end the discussion of genetics and your world without some final comments on the ethical issues that genetic advances raise. In this chapter, you find out how genetics has been misunderstood, misinterpreted, and misused to cause people harm based on their racial, ethnic, or socioeconomic status. The rapidly growing field of genetics is contributing to ideas about how modern humans can mold the future of their offspring, so this chapter dispels the myth of the designer baby. You discover how information you give out and receive can be used for and against you. Finally, you gain a better understanding of the next generation of studies based on the Human Genome Project and the ethical issues that mapping human genetic diversity will bring up.

Profiling Genetic Racism

One of the biggest hot button issues of all time has to be *eugenics*. In a nutshell, eugenics is the idea that humans should practice selective reproduction in an effort to "improve" the species. If you read Chapter 19, which explains how organisms can be genetically engineered, you probably already have some idea of what eugenics in the modern age may entail (transgenic, made-to-order babies, perhaps?). Historically, the most blatant examples of eugenics are genocidal activities the world over. (Perhaps the most infamous example occurred in Nazi Germany during the 1930s and 1940s.)

The story of eugenics begins with the otherwise laudable Francis Galton, who coined the term in 1883. (Galton is best remembered for his contribution to law enforcement: He invented the process used to identify persons by their fingerprints. Check out Chapter 18 for more on the genetic version of fingerprinting.) In direct and vocal opposition to the U.S. Constitution, Galton was quite sure that all men were *not* created equal (I emphasize here that he was particularly fixated on men; women were of no consequence in his day). Instead, Galton believed that some men were quite superior to others. To this end, he attempted to prove that "genius" is inherited. The view that superior intelligence is heritable is still widely held despite abundant evidence to the contrary. For example, twin studies conducted as far back as the 1930s show that genetically identical persons are not intellectually identical.

Galton, who was one of Charles Darwin's cousins, gave eugenics its name, but his ideas weren't unique or revolutionary. During the early 20th century, as understanding of Mendelian genetics (see Chapter 3) gathered steam, many people viewed eugenics as a highly admirable field of study. Charles Davenport was one such person. Davenport holds the dubious distinction of being the father of the American eugenics movement (one of his eugenics texts is subtitled "The science of human improvement by better breeding"). The basis of Davenport's idea is that "degenerate" people shouldn't reproduce. This notion arose from something called *degeneracy theory,* which posits that "unfit" humans acquire certain undesirable traits because of "bad environments" and then pass on these traits genetically. To these eugenicists, unfit included "shiftlessness," "feeblemindedness," and poverty, among other things.

While the British, including Galton, advocated perpetuating good breeding (along with wealth and privilege), many American eugenicists focused their attention on preventing *cacogenics,* which is the erosion of genetic quality. Therefore, they advocated forcibly sterilizing people judged undesirable or merely inconvenient. Shockingly, the forcible sterilization laws of this era have never been overturned, and until the 1970s, it was still common practice to sterilize mentally ill persons without their consent — an estimated 60,000 people in the United States suffered this atrocity. Some societies have taken this sick idea a step further and *murdered* the "unfit" in an effort to remove them and their genes permanently.

Sadly, violent forms of eugenics, such as genocide, rape, and forced sterilization, are still advocated and practiced all over the world. But not all forms of eugenics are as easy to recognize as these extreme examples. To some degree, eugenics lies at the heart of most of the other ethical quandaries I address in this chapter. In addition, it only requires a little imagination to see how gene therapy (Chapter 16), gene transfer (Chapter 19), or DNA fingerprinting (Chapter 18) can be abused to advance the cause of eugenics.

Ordering Up Designer Babies

One of the more contentious issues with a root in eugenics stems from a combination of prenatal diagnosis and the fantasy of the perfect child to create a truly extreme makeover — designer babies. In theory, a designer baby may be made-to-order according to a parent's desire for a particular sex, hair and eye color, and maybe even athletic ability.

The myth of designer babies

The term *designer baby* gets tossed around quite a bit these days. In essence, the term is associated with genetically made-to-order offspring. As of this writing, neither the technology nor sufficient knowledge of the human genome exists to make the designer baby a reality.

The fantasy of the designer baby, like cloning (see Chapter 20), rests on the fallacy of *biological determinism* (which, by the way, is what eugenics bases some of its lies on, too; jump back to "Profiling Genetic Racism" to find out about eugenics). Biological determinism assumes that genes are expressed in precise, repeatable ways — in other words, genetics is identity is genetics. However, this assumption isn't true. Gene expression is highly dependent on environment, among other things (see Chapter 10 for more details about how gene expression works).

Furthermore, the *in vitro* fertilization process that plays a role in current-day applications of the science in question (see the next section) is a very dicey and difficult process at best — just ask any couple who's gone through it in an effort to get pregnant. *In vitro* procedures are extremely expensive, invasive, and painful, and women must take large quantities of strong and potentially dangerous fertility medications to produce a sufficient number of eggs. And in the end, the majority of fertilizations don't result in pregnancies.

The reality of the science:
Prenatal diagnosis

So where does the myth of designer babies come from? Using procedures similar to those leading up to cloning (covered in Chapter 20), *preimplantation genetic diagnosis,* or PGD, is performed before a fertilized egg implants in the womb. Although it's true that PGD opens the remote possibility of creating transgenic humans using the same technology used to create transgenic animals (see Chapter 19 for details), the likelihood of PGD becoming commonplace is extremely remote.

The process of PGD is technologically complicated. First, unfertilized eggs are harvested from a female donor. *In vitro fertilization* (the process to produce the so-called test-tube baby) is performed, and then the fertilized eggs are tested for specific gene mutations or other genetic variations. In a few rare cases, desperate parents have created embryos this way specifically to look for genetic compatibility with preexisting offspring — the plan being to conceive a sibling who can provide stem cells or bone marrow to save the life of a living sibling suffering from an otherwise untreatable disease. Saving the lives of living children is undoubtedly a laudable goal; the problem arises with what's done with the fertilized eggs that don't meet the desired criteria (if, for example, they don't have the desired tissue match). Even if inserted into the mother's uterus, the vast majority of these fertilized eggs would never implant and thus would not survive. Although lack of implantation is also true when conception occurs naturally, it's still a very tough call to decide the fate of extra embryos. Options include donation to other couples, donation for research purposes, or destruction.

PGD and other forms of prenatal diagnosis allow parents the choice to prevent, alleviate, or reduce suffering (their own or someone else's). But like deciding the fate of extra embryos, this is very deep and muddied water. Without getting too philosophical, suffering is a highly personal experience; that is, what constitutes suffering to one person may look relatively okay to someone else. One example of relative suffering that comes up a lot is hereditary deafness. If a deaf couple chooses prenatal diagnosis, what's the most desirable outcome? On one hand, a deaf child shares the worldview of his or her parents. On the other hand, a hearing child fits into the world of nondeaf people more easily. By now, you see how complex the issues surrounding prenatal diagnosis are. It seems clear that right answers, if there are any, will be very hard to come by.

Who Knows? Getting Informed Consent

Informed consent is a sticky ethical and legal issue. Basically, the idea is that a person can only truly make a decision about having a procedure when he or she is fully apprised of all the facts, risks, and benefits. Informed consent can only be given by the person receiving the procedure or by that person's legal guardian. Generally, guardianship is established in cases where the recipient of the procedure is too young to make decisions for him or herself or is mentally incapacitated in some way; presumably, guardians have the best interests of their wards at heart.

Three major issues exist in the debate over informed consent:

- ✔ Genetic testing can be carried out on embryos, the deceased, and samples obtained from anyone during the simplest of medical procedures.

- ✔ Experimental genetic treatments (that is, gene therapies; see Chapter 16) have, by their very nature, unpredictable outcomes, making risk difficult to quantify to prospective participants.

- ✔ After tissue samples are obtained and genetic profiling is done, information storage and privacy assurance could be problematic.

Placing restrictions on genetic testing

Genetic testing in the forms of DNA fingerprinting, SNP analysis (see Chapter 18), and gene sequencing (see Chapter 11), among others, is now routine, fast, and relatively cheap. The testing can glean massive amounts of information — from an individual's sex to his or her racial and ethnic makeup — from even a very tiny sample of tissue. The procedure can also detect the presence of mutations for inherited disorders. But given that your DNA has so much personal information stored in it, shouldn't you have complete control over whether you're tested? The answer to this question is becoming more and more contentious as the definitions of, and limits to, informed consent are explored. The rights of persons both living and dead are at stake.

For example, the descendents of Thomas Jefferson consented to genetic testing in 1998 to settle a long-standing controversy about Jefferson's relationship with one of his slaves, Sally Hemings (see Chapter 18 for the full story). In the Jefferson case, the matter was more than just academic curiosity because the right to burial in the family cemetery at Monticello was at stake.

The issue of informed consent, or lack thereof, is complicated by the ability to store tissue for long periods of time. In some cases, patients or their guardians gave informed consent for certain tests but didn't include tests that hadn't yet been developed. Some institutions routinely practice long-term tissue storage, making informed consent a frequent point of contention. For example, a children's hospital in Britain was taken to task over storage of organs that were obtained during autopsies but weren't returned for internment with the rest of the body. Parents of the affected deceased gave consent for the autopsies but not for the retention of tissues.

Biologists also use stored tissue to create *cell-lines,* living tissues that grow in culture tubes for research purposes. The original cell donors are often dead, usually from the disease under study. Cell-lines aren't that hard to make and maintain (if you know what you're doing), but the creation of cell-lines raises the question of whether the original donor has ownership rights to cells descended from his or her tissue. Cell-lines sometimes result in patents for lucrative treatments; should donors or their heirs get a royalty? (A court decision in California said no.)

Moreover, some cell-lines are developed using unfertilized human eggs, creating another ethical quandary. One type of stem cell research fuses an egg (that has had its nucleus removed) with an adult somatic cell in an attempt to create a stem cell-line that's matched with the somatic cell donor's tissue. This sort of research requires huge numbers of human eggs. In a controversial move, the New York Stem Cell Foundation's ethics committee voted to allow women to be paid for their eggs. Though limiting payments to compensation for "time and burden," the decision increases worries about both placing egg donors' health at risk and the commercialization of human body parts.

Practicing safe genetic treatment

If you've ever had to sign a consent for treatment form, you know it can be a sobering experience. Almost all such forms include some phrase that communicates the possibility of death. With a gulp, most of us sign off and hope for the best. For routine procedures and treatments, our faith is usually repaid with survival. Experimental treatments are harder to gauge, though, and fully informing someone about possible outcomes is very difficult.

The 1999 case of Jesse Gelsinger (covered in Chapter 16) brought the problem of informed consent and experimental treatment into a glaring, harsh light. Jesse died after receiving an experimental treatment for a hereditary disorder that, by itself, wasn't likely to kill him. His treatment took place as part of a clinical trial designed to assess the effects of a particular therapy in relatively healthy patients and to work out any difficulties before initiating treatments on patients for whom the disease would, without a doubt, be fatal (in this case, infants homozygous for the allele). What researchers knew about all the possible outcomes and what the Gelsinger family was told before treatment began is debatable.

Almost every article on gene therapy published since the Gelsinger case makes mention of it. In fact, most researchers in the field divide the development of gene therapies into two categories: before and after Gelsinger. Sadly, Gelsinger's death probably contributed very little to the broader understanding of gene therapy. Instead, the impacts of the Gelsinger case are that clinical trials are now harder to initiate, criteria for patient inclusion and exclusion are heightened, and disclosure and reporting requirements are far more stringent. These changes are basically a double-edged sword: New regulations protect patients' rights and simultaneously decrease the likelihood that researchers will develop treatments to help those who desperately need them. Like so many ethical issues, a safe and effective solution may prove elusive.

Keeping it private

Another issue in the informed consent debate relates to privacy. When genetic tests are conducted, the data recorded often includes detailed medical histories and other personal information, all of which aids researchers or physicians in the interpretation of the genetic data obtained. So far, so good. But what happens to all that information? Who sees it? Where's it stored? And for how long?

Privacy is a big deal, particularly in American culture. Laws exist to protect one's private medical information, financial status, and juvenile criminal records (if any). Individuals are protected from unwarranted searches and surveillance, and they have the right to exclude unwanted persons from their private property. Genetic information is likely to fall under existing medical privacy laws, but there's one twist: Genetic information contains an element of the future, not just the past.

When you carry a mutation for susceptibility to breast cancer, you have a greater likelihood of developing breast cancer than someone who doesn't have the allele (see Chapter 14). A breast cancer allele doesn't guarantee you'll develop the cancer, though; it just increases the probability. If you were to be tested for the breast cancer allele and found to have it, that information would become part of your medical record. Besides your doctor and appropriate medical personnel, who might learn about your condition? Your insurance company, that's who. So far, situations like this haven't presented a big problem because few people have had genetic tests. Many genetic tests are often expensive and aren't part of routine healthcare, but as technology advances and gets cheaper (like microarrays; see Chapter 23), genetic testing is likely to become more common. And that shift may be both a blessing and a curse.

As a patient, knowing that you have a genetic mutation is a really good thing, because the condition may be treatable, or an early detection screening may help you prevent more serious complications. For example, cancers that are caught early have far better prognoses than those diagnosed in later stages. However, knowing about a genetic mutation may give insurance companies the chance to issue or cancel policies, thus unfairly limiting your access to healthcare or employment. Sadly, at least one employer has been caught attempting to test workers for genetic predispositions to certain injuries (in this case, carpal tunnel syndrome, a repetitive stress injury to the hands and arms) without the employees' knowledge — clearly, a violation of informed consent.

Genetic privacy issues also feed into the controversies surrounding the Human Genome Project and efforts to characterize human population genetics. Critics fear that if certain mutations or health problems are genetically linked to groups of people, discrimination and bias will result. Fortunately, lawmakers are taking steps to protect you and your genetic privacy. In 2008, former U.S. President George W. Bush signed the Genetic Information Nondiscrimination Act (GINA), which prohibits both health insurance companies and employers from discriminating against someone based on information from genetic testing. In addition, most state legislatures have enacted similar laws.

Genetic Property Rights

According to U.S. law, a patent gives the patent owner exclusive rights to manufacture and sell his or her invention for a certain length of time (usually 20 years). That may not sound like a big deal, but what makes patents scary is that companies are patenting *genes* — DNA sequences that hold the instructions for life. And it's not just any genes, either. They're patenting *your* genes.

Patents are granted to *inventors,* but the people (or companies) holding gene patents didn't invent the genes that naturally occur in living organisms. According to most legal experts, genes are "unpatentable products of nature." Yet so far, American and European patenting authorities have viewed genes in the same legal light as manmade chemicals. Generally, patent-holding companies sequence the genes and convert them to another form called cDNA (*c* means complementary; see Chapter 16 for coverage of translation). Then they seek a patent on the cDNA rather than the gene itself. Another approach to the patenting process is that the company discovers a gene (or a disease-causing version of it) and then invents products such as diagnostic tests that have something to do with the gene.

Just how a company can own and exercise exclusive rights over your genes is a little hard to understand. An example of how gene patenting works comes from the invention of the process of PCR (see Chapter 22 for the whole tale). The process uses an enzyme that's produced by a very special sort of bacteria. The gene that codes for that enzyme (called *Taq polymerase*) is easily moved into other bacteria, such as *E. coli,* using recombinant DNA techniques (explained in Chapter 16). This means that *E. coli* can produce the enzyme that can then be used to run PCR. But if any other geneticist uses that gene to make Taq polymerase, a royalty must be paid to the company that patented it. Not surprisingly, this company is now the biggest manufacturer of Taq in the world, raking in profits in the billions of dollars.

Here are examples of how ugly the gene-patenting game can get:

- In 2001, an American company got a European patent for *BRCA1*, one of the breast cancer genes (see Chapter 14 for a full description of this mutation). A mutation in this gene can lead to cancer, and presumably, no one would want to purchase a case of breast cancer. So why patent it? Because the company holding the patent can charge large sums to test people to determine whether they carry the mutation.

- A large drug company holds a patent on a gene test that can determine whether the company's product will work for certain persons. The company refuses to actually develop the test or let anyone else have a crack at it because doing so may reduce sales of the medication in question.

- Companies patent disease-causing bacteria and viral genes for the same reasons — to block diagnosis and treatment — until a hefty licensing fee has been paid.

Such use of genetic patents impedes both research to combat disease and access to healthcare. Because of these kinds of manipulations, gene patents are beginning to meet with strong and vocal opposition.

Gene-patenting policies may endanger your health in other ways as well.

- When commercial outfits get genetic information, they treat it as their personal property. Therefore, they don't always report gene sequences and experiment results in the appropriate scientific literature (thus avoiding review and verification by experts in the field). To market their products, these companies must go through the regulatory process mandated by the government to ensure consumer safety, but that regulatory review process has suffered noticeable shortcomings of late — particularly when products are allowed to pass muster while some conflict of interest is at work (think stock options, as was the case in shake-ups at the U.S. National Institutes of Health in 2005).

- Sadly, universities have gotten in on the act. In one instance, the search for the genes responsible for autism was held up because several universities refused to share information with, of all people, the parents of autistic children. Each university wanted to be the first to (you guessed it) patent the "autism gene." As a result, an independent foundation was established to create a public repository for genetic information about autism, because such actions are a direct assault on the openness of the scientific research itself.

The days of gene patents may be numbered, however. In 2009, cancer patient Genae Girard, along with four other patients, genetics researchers, and others, filed suit against a company that owns the patent on two genes associated with breast cancer risk. If their challenge is successful, restrictions on testing and other sorts of research could be lifted, paving the way to more open exchange of knowledge and lowering the costs of testing.

Part V
The Part of Tens

"So, Bateson—how's the work in genetics and plant hybridization coming?"

In this part . . .

Genetics is equal parts great history and amazing future. The discoveries of the past depended on the genius of many individuals. Likewise, the marvels of the future will be shaped by teams of researchers and entrepreneurs.

This part exposes you to genetics' past and allows you to glimpse its future as well. I introduce you to the ten most important people and events that shaped what genetics is today, and I explain the next big things (or ten of them, at least) on the genetics horizon. Finally, I shake things up with ten hard-to-believe (and all true!) genetics stories.

Chapter 22

Ten Defining Events in Genetics

Many milestones define the history of genetics. This chapter focuses on nine that I don't cover in other chapters of the book and one that I do (the Human Genome Project is so important that I cover it in Chapter 8 and here, too). The events listed here appear roughly in order of historical occurrence.

The Publication of Darwin's "The Origin of Species"

Earthquakes have aftershocks — little mini-earthquakes that rattle around after the main quake. Events in history sometimes cause aftershocks, too. The publication of one man's life's work is such an event. From the moment it hit the shelves in 1856, Charles Darwin's *The Origin of Species* was deeply controversial (and still is).

The basis of evolution is elegantly simple: Individual organisms vary in their ability to survive and reproduce. For example, a sudden cold snap occurs, and most individuals of a certain bird species die because they can't tolerate the rapid drop in temperature. But individuals of the same species that can tolerate the unexpected freeze survive and reproduce. As long as the ability to deal with rapid temperature drops is heritable, the trait is passed to future generations, and more and more individuals inherit it. When groups of individuals are isolated from each other, they wind up being subjected to different sorts of events (such as weather patterns). After many, many years, stepwise changes in the kinds of traits that individuals inherit based on events like a sudden freeze accumulate to the point that populations with common ancestors become separate species.

Darwin concluded that all life on earth is related by inheritance in this fashion and thus has a common origin. Darwin arrived at his conclusions after years of studying plants and animals all over the world. What he lacked was a convincing explanation for how individuals inherit advantageous traits. Yet the explanation was literally at his fingertips. Gregor Mendel figured out the laws of inheritance at about the same time that Darwin was working on his book (see Chapter 3). Apparently, Darwin failed to read Mendel's paper — he scrawled notes on the papers immediately preceding and following Mendel's paper but left Mendel's unmarked. Darwin's copious notes show no evidence that he was even aware of Mendel's work.

Even without knowledge of how inheritance works, Darwin accurately summarized three principles that are confirmed by genetics:

- ✓ **Variation is random and unpredictable.** Studies of mutation confirm this principle (see Chapter 13).

- ✓ **Variation is *heritable* (it can be passed on from one generation to the next).** Mendel's own research — and thousands of studies over the past century — confirms heritability. With DNA fingerprinting, heritable genetic variation can be traced directly from parent to offspring (see Chapter 18 for how paternity tests use heritable genetic markers to determine which male fathered which child).

- ✓ **Variation changes in frequency over the course of time.** The Hardy-Weinberg principle formalized this concept in the form of population genetics in the early 1900s (see Chapter 17). Since the 1970s, genetic studies using DNA sequencing (along with other methods) have confirmed that genetic variation within populations changes because of mutation, accidents, and geographic isolation, to name only a few causes.

Regardless of how you view it, the publication of Darwin's *The Origin of Species* is pivotal in the history of genetics. If no genetic variation existed, all life on earth would be precisely identical. Variation gives the world its rich texture and complexity, and it's what makes you wonderfully unique.

The Rediscovery of Mendel's Work

In 1866, Gregor Mendel wrote a summary of the results of his gardening experiments with peas (which I detail in Chapter 3). His work was published in the scientific journal *Versuche Pflanzen Hybriden,* where it gathered dust for nearly 40 years. Although Mendel wasn't big on self-promotion, he sent copies of his paper to two well-known scientists of his time. One copy remains missing; the other was found in what amounts to an unopened envelope — the pages were never cut. (Old printing practices resulted in pages being folded together; the only way to read the paper was to cut the pages apart.) Thus, despite the fact that his findings were published and distributed (though limitedly), his peers didn't grasp the magnitude of Mendel's discovery.

Mendel's work went unnoticed until three botanists — Hugo de Vries, Erich von Tschermak, and Carl Correns — all reinvented Mendel's wheel, so to speak. These three men conducted experiments that were very similar to Mendel's. Their conclusions were identical — all three "discovered" the laws of heredity. De Vries found Mendel's work referenced in a paper published in 1881. (De Vries coined the term *mutation,* by the way.) The author of the 1881 paper, a man by the unfortunate name of Focke, summarized Mendel's findings but didn't have a clue as to their significance. De Vries correctly interpreted Mendel's work and cited it in his own paper, which was published in 1900. Shortly thereafter, Tschermak and Correns also discovered Mendel's publication through de Vries's published works and indicated that their own independent findings confirmed Mendel's conclusions as well.

William Bateson is perhaps the great hero of this story. He was already incredibly influential by the time he read de Vries's paper citing Mendel, and unlike many around him, he recognized that Mendel's laws of inheritance were revolutionary and absolutely correct. Bateson became an ardent voice spreading the word. He coined the terms *genetics, allele* (shortened from the original *allelomorph*), *homozygote,* and *heterozygote.* Bateson was also responsible for the discovery of linkage (see Chapter 4), which was experimentally confirmed later by Morgan and Bridges.

The Transforming Principle

Frederick Griffith wasn't working to discover DNA. The year was 1928, and the memory of the deadly flu epidemic of 1918 was still fresh in everyone's mind. Griffith was studying pneumonia in an effort to prevent future epidemics. He was particularly interested in why some strains of bacteria cause illness and other seemingly identical strains do not. To get to the bottom of the issue, he conducted a series of experiments using two strains of the same species of bacteria, *Streptococcus pneumonia.* The two strains looked very different when grown in a Petri dish, because one grew a smooth carpet and the other a lumpy one (he called it "rough"). When Griffith injected smooth bacteria into mice, they died; rough bacteria, on the other hand, were harmless.

To figure out why one strain of bacteria was deadly and the other wasn't, Griffith conducted a series of experiments. He injected some mice with heat-killed smooth bacteria (which turned out to be harmless) and others with heat-killed smooth in combination with living rough bacteria. This combo proved deadly to the mice. Griffith quickly figured out that something in the smooth bacteria *transformed* rough bacteria into a killer. But what? For lack of anything better, he called the responsible factor the *transforming principle* (which now sounds like a good title for a diet book).

Oswald Avery, Maclyn McCarty, and Colin MacLeod teamed up in the 1940s to discover that Griffith's transforming principle was actually DNA. This trio made the discovery by a dogged process of elimination. They showed that fats and proteins don't do the trick; only the DNA of smooth bacteria provides live rough bacteria with the needed ingredient to get nasty. Their results were published in 1944, and like Mendel's work nearly a century before, their findings were largely rejected.

It wasn't until Erwin Chargaff came along that the transforming principle started to get the appreciation it deserved. Chargaff was so impressed that he changed his entire research focus to DNA. Chargaff eventually determined the ratios of bases in DNA that helped lead to Watson and Crick's momentous discovery of DNA's double helix structure (flip to Chapter 6 for all the details).

The Discovery of Jumping Genes

By all accounts, Barbara McClintock was both brilliant and a little odd; a friend once described her as "not fooled or foolable." McClintock was unorthodox in both her research and her outlook as she lived and worked alone for most of her life. Her career began in the early 1930s and took her into a man's world — very few women worked in the sciences in her day.

In 1931, McClintock collaborated with another woman, Harriet Creighton, to demonstrate that genes are located on chromosomes. This fact sounds so self-evident now, but back then, it was a revolutionary idea. Creighton and McClintock showed that corn chromosomes recombine during meiosis (see Chapter 2 for the scoop on meiosis). By tracking the inheritance of various traits, they figured out which genes were getting moved during translocation events (see Chapter 15). *Translocations* hook up chunks of chromosomes in places where they don't belong. Chromosomes with translocations look very different from normal chromosomes, making it easy to track their inheritance. By linking physical traits to certain parts of one odd-looking chromosome, Creighton and McClintock demonstrated that crossover events between chromosomes move genes from one chromosome to another.

McClintock's contribution to genetics goes beyond locating genes on chromosomes, though. She also discovered traveling bits of DNA, sometimes known as jumping genes (see Chapter 11 for more). In 1948, McClintock, working independently, published her results demonstrating that certain genes of corn could hop around from one chromosome to another *without* translocation. Her announcement triggered little reaction at first. It's not that people thought McClintock was wrong; she was just so far ahead of the curve that her fellow geneticists couldn't comprehend her findings. Alfred Sturtevant (who was responsible for the discovery of gene mapping) once said, "I didn't understand one word she said, but if she says it is so, it must be so!"

It took nearly 40 years before the genetics world caught up with Barbara McClintock and awarded her the Nobel Prize in Physiology or Medicine in 1983. By then, jumping genes had been discovered in many organisms (including humans). Feisty to the end, this grand dame of genetics passed away in 1992 at the age of 90.

The Birth of DNA Sequencing

So many events in the history of genetics lay a foundation for other events to follow. Federick Sanger's invention of chain-reaction DNA sequencing (which I cover in Chapter 8) is one of those foundational events. In 1980, Sanger shared his second Nobel Prize (in Chemistry) with Walter Gilbert for their work on DNA. Sanger had already earned a Nobel Prize in Chemistry in 1958 for his pioneering work on the structure of the protein insulin. (*Insulin* is produced by your pancreas and regulates blood sugar; its absence results in diabetes.)

Sanger figured out the entire process used for DNA sequencing. Every single genetics project that has anything to do with DNA uses Sanger's method. *Chain-reaction sequencing,* as Sanger's method is called, uses the same mechanics as replication in your cells (see Chapter 7 for a rundown of replication). Sanger figured out that he could control the DNA building process by snipping off one oxygen molecule from the building blocks of DNA. The resulting method allowed identification of every base, in order, along a DNA strand, sparking a revolution in the understanding of how your genes work. This process is responsible for the Human Genome Project, DNA fingerprinting (see Chapter 18), genetic engineering (see Chapter 19), and gene therapy (see Chapter 16).

The Invention of PCR

In 1985, while driving along a California highway in the middle of the night, Kary Mullis had a brainstorm about how to carry out DNA replication in a tube (see Chapter 7 for the scoop on replication). His idea led to the invention of *polymerase chain reaction* (PCR), a pivotal point in the history of genetics.

I detail the entire process of how PCR is used in DNA fingerprinting in Chapter 18. In essence, PCR acts like a copier for DNA. Even the tiniest snippet of DNA can be copied. This concept is important because, so far, technology isn't sophisticated enough to examine one DNA molecule at a time. Scientists need many copies of the same molecule before enough is present for them to detect and study. Without PCR, large amounts of DNA are needed to generate a DNA fingerprint, but at many crime scenes, only tiny amounts of DNA are present. PCR is the powerful tool that every crime lab in the country now uses to detect the DNA left behind at crime scenes and to generate DNA fingerprints.

Mullis's bright idea turned into a billion dollar industry. Although he reportedly was paid a paltry $10,000 for his invention, he received the Nobel Prize for Chemistry in 1993 (a sort of consolation prize).

The Development of Recombinant DNA Technology

In 1970, Hamilton O. Smith discovered *restriction enzymes,* which act as chemical cleavers to chop DNA into pieces at very specific sequences. As part of other research, Smith put bacteria and a bacteria-attacking virus together. The bacteria didn't go down without a fight — instead, it produced an enzyme that chopped the viral DNA into pieces, effectively destroying the invading virus altogether. Smith determined that the enzyme, now known as *HindII* (named for the bacteria *Haemophilus influenzae Rd*), cuts DNA every time it finds certain bases all in a row and cuts between the same two bases every time.

This fortuitous (and completely accidental!) discovery was just what was needed to spark a revolution in the study of DNA. Some restriction enzymes make offset cuts in DNA, leaving single-stranded ends. The single-strand bits of DNA allow geneticists to "cut-and-paste" pieces of DNA together in novel ways, forming the entire basis of what's now known as *recombinant DNA technology.*

Gene therapy (see Chapter 16), the creation of genetically engineered organisms (see Chapter 19), and practically every other advance in the field of genetics these days all depend on the ability to cut DNA into pieces without disabling the genes and then to put the genes into new places — a feat made possible thanks to restriction enzymes.

Today, researchers use thousands of restriction enzymes to help map genes on chromosomes, determine the function of genes, and manipulate DNA for diagnosis and treatment of disease. Smith shared the Nobel Prize in Physiology or Medicine in 1978 with two other geneticists, Dan Nathans and Werner Arber, for their joint contributions to the discovery of restriction enzymes.

The Invention of DNA Fingerprinting

Sir Alec Jeffreys has put thousands of wrongdoers behind bars. Almost single-handedly, he's also set hundreds of innocent people free from prison. Not bad for a guy who spends most of his time in the genetics lab.

Jeffreys invented DNA fingerprinting in 1985. By examining the patterns made by human DNA after it was diced up by restriction enzymes, Jeffreys realized that every person's DNA produces a slightly different number of various sized fragments (which number in the thousands).

Jeffreys's invention has seen a number of refinements since its inception. PCR and the use of STRs (*short tandem repeats;* see Chapter 18) have replaced the use of restriction enzymes. Modern methods of DNA fingerprinting are highly repeatable and extremely accurate, meaning that a DNA fingerprint can be stored much like a fingerprint impression from your fingertip. More than 100 laboratories in the United States alone now make use of the methods that Jeffreys pioneered. The information that these labs generate is housed in a huge database hosted by the FBI, granting any police department quick access to data that can help match criminals to crimes.

In 1994, Queen Elizabeth II knighted Jeffreys for his contributions to law enforcement and his accomplishments in genetics.

The Explanation of Developmental Genetics

As I explain in Chapter 8, every cell in your body has a full set of genetic instructions to make all of you. The master plan of how an entire organism is built from genetic instructions remained a mystery until 1980, when Christiane Nüsslein-Volhard and Eric Wieschaus identified the genes that control the whole body plan during fly development.

Fruit flies and other insects are constructed of interlocking pieces, or segments. A group of genes (collectively called *segmentation genes*) tells the cells which body segments go where. These genes, along with others, give directions like top and bottom and front and back, as well as the order of body regions in between. Nüsslein-Volhard and Wieschaus made their discovery by mutating genes and looking for the effects of the "broken" genes. When segmentation genes get mutated, the fly ends up lacking whole sections of important body parts or certain pairs of organs.

A different set of genes (called *homeotic genes*) controls the placement of all the fly's organs and appendages, such as wings, legs, eyes, and so on. One such gene is *eyeless.* Contrary to what would seem logical, *eyeless* actually codes for normal eye development. Using the same recombinant DNA techniques made possible by restriction enzymes (see the section "The Development of Recombinant DNA Technology" earlier in the chapter), Nüsslein-Volhard and Wieschaus moved *eyeless* to different chromosomes where it could be turned on in cells in which it was normally turned off. The resulting flies grew eyes in all sorts of strange locations — on their wings, legs, butts, you name it. This research showed that, working together, segmentation and homeotic genes put all the parts in all the right places. Humans have versions of these genes, too; your body-plan genes were discovered by comparing fruit fly genes to human DNA (see Chapter 8 for how the genomes of organisms affect you).

The Work of Francis Collins and the Human Genome Project

In 1989, Francis Collins and Lap-Chee Tsui identified the single gene responsible for cystic fibrosis. The very next year, the Human Genome Project (HGP) officially got underway. A double-doctor (that is, a doc with an MD and PhD), Collins later replaced James Watson as the head of the National Human Genome Research Institute in the United States and supervised the race to sequence the entire human genome from start to finish. In 2009, Collins became the director of the U.S. National Institutes of Health.

Collins is one of the true heroes of modern genetics. He kept the HGP ahead of schedule and under budget. He continues to champion the right to free access to all the HGP data, making him a courageous opponent of gene patents and other practices that restrict access to discovery and healthcare, and he's a staunch defender of genetic privacy (see Chapter 21 for more on these subjects). Although the human genome is still bits and pieces away from being completely sequenced, the project wouldn't have been a success without the tireless work of Collins, who's still an active gene hunter. His lab is now searching out the genes responsible for adult-onset diabetes.

Chapter 23

Ten Hot Issues in Genetics

Genetics is a field that grows and changes with every passing day. The hottest journals in the field *(Nature* and *Science)* are full of new discoveries each and every week. This chapter shines the spotlight on ten of the hottest topics and next big things in this ever-changing scientific landscape.

Personalized Medicine

The fourth biggest cause of death in the United States is adverse reactions to medications. Up to 100,000 people die each year from something that's meant to help them. Why? The tool scientists use to answer that question is *pharmacogenomics,* the analysis of the human genome and heredity to determine how drugs work in individual people. The idea is that the reason certain people have adverse reactions to certain drugs and others don't lies somewhere in their DNA. If researchers could develop a simple test to detect these DNA differences, doctors would never prescribe the wrong drugs in the first place. (Oddly, this idea sometimes doesn't go over well with drug companies; for more on the connection between the two, check out Chapter 21.) The overarching goal of personalized medicine is a new brand of care that can be designed to fit the unique genetic makeup of each individual patient.

That's the good news. The bad news is that nobody knows how many genes are involved in diseases, and many genes can cause the *same* disease. Not only that, but *epigenetics* (flip to Chapter 4 to find out more about epigenetics) further complicates matters by turning genes on and off in unexpected ways. All this adds up to more confusion and fewer genetics-based treatments, meaning that the promises of personalized medicine may wind up being slow, or impossible, to realize.

Stem Cell Research

Stem cells may hold the key to curing brain and spinal cord injuries. They may be part of the cure for cancer. These little wonders may be *the* magic bullet to solving all sorts of medical problems, but they're at the center of controversies so big that their potential remains untested.

Stem cells are hot research topics because they're totipotent. *Totipotence* means that stem cells can turn into any kind of tissue, from brain to muscle to bone, just to name a few. Not too surprisingly, stem cells are what undifferentiated embryos are made of; that is, a fertilized egg, shortly after it starts dividing, is composed entirely of stem cells. At a certain point during development, all the cells get their assignments, and totipotence is long gone (except for DNA, which retains surprising flexibility — DNA's totipotence is what allows cloning to work; see Chapter 20).

You've probably guessed (or already knew) that the source of stem cells for research is embryonic tissue — and therein lies the rub. As of this writing, researchers haven't found a way to harvest stem cells without sacrificing the embryo in the process. They can collect stem cells from adults (from various places, including blood), but adult stem cells lack some of the totipotent potential of embryonic cells and are present in very low numbers, which makes using adult stem cells problematic. Nonetheless, adult stem cells may work better than embryonic ones for therapeutic purposes, because researchers can harvest them from a patient, modify them, and return them to the patient, eliminating the chance of tissue rejection. (For the lowdown on gene therapy, see Chapter 16.) A potential compromise may be collecting the cells from an umbilical cord after a child is born; these cells are even better than adult stem cells. Stem cells in one form or another may yet find their way into modern medicine, but for now, moral and ethical opposition to the use of embryonic cells stymies stem cell research because most of it depends on the use of embryonic tissues.

Aging Genes

Aging is not for the timid. Skin sags, hair turns gray, joints hurt. Sounds like fun, doesn't it? The effects may be obvious, but the process of *senescence* (the fancy term for aging) is still quite a mystery. Scientists know that the ends of your chromosomes (called *telomeres*) sometimes get shorter as you get older (see Chapter 7), but they aren't sure that those changes are what make old folks old. What is known is that when telomeres get too short, cells die, and cell death is clearly part of the aging process.

The enzyme that can prevent telomeres from shortening, *telomerase* (see Chapter 7), seems an obvious target for anti-aging research. Cells that have active telomerase don't die because of shortened telomeres. For instance, cancer cells often have active telomerase when normal cells don't; telomerase activity contributes to the unwanted longevity that cancer cells enjoy (flip to Chapter 14 for the details). If geneticists can get a handle on telomerase — turning it on where it's wanted without causing cancer — aging may become controllable.

In addition, geneticists have discovered that old cells perk up when put in the company of younger cells. This finding indicates that cells have plenty of capacity to regenerate themselves — they just need a little incentive. Another recent study suggests that calorie restriction in a person's diet also helps defer the effects of aging. Researchers found that when mice were put on a calorie-restricted diet, a gene kicked in to slow programmed cell death (called *apoptosis;* see Chapter 14).

New information on how to prevent aging is in high demand. If keeping young turns out to be as simple as spending time with younger people and eating less, aging may be a lot more fun than it seems.

Proteomics

Genomics, the study of whole genomes, will soon have to make room for the next big thing: *proteomics,* the study of all the proteins an organism makes. Proteins do all the work in your body. They carry out all the functions that genes encode, so when a gene mutation occurs, the protein is what winds up being altered (or goes missing altogether). Given the link between genes and proteins, the study of proteins may end up telling researchers more about genes than the genes themselves!

Proteins are three-dimensional (see Chapter 9 for an explanation). Proteins not only get folded into complex shapes but also get hooked up with other proteins and decorated with other elements such as metals. (See Chapter 9 for more on how proteins are modified from plain amino acid chains to get gussied up to do their jobs.) Currently, scientists can't just look at a protein and tell what its function is. If it's possible to decode them, though, proteins may be a big deal in the fields of medical drugs and treatments, because medications act upon the proteins in your system.

Cataloging all the proteins in your proteome hasn't been easy, because researchers have to sample every tissue to find them all. Nonetheless, the rewards of discovering new drugs and treatments for previously untreatable diseases may make the effort worthwhile. Like personalized medicine, however, proteomics hasn't made a big splash in clinical settings just yet — complexities and technological setbacks have slowed progress.

Bioinformatics

You live in the information age, with practically everything you need at your fingertips. But where genetics is concerned, it's the information overflow age — thousands and thousands of DNA sequences, gobs of proteins, tons of data. It's hard to know where to start or how to sort through the mountains of chatter to get to the real messages. Never fear! *Bioinformatics For Dummies* is here! (I'm not kidding. It's a real *For Dummies* title. For specifics, check out www.dummies.com.)

Bioinformatics is the process of using a computer to sort through massive biological databases. Anyone with an Internet connection can access these databases with the click of a mouse (surf to www.ncbi.nlm.nih.gov to reach the National Center for Biotechnology Information). Hop online and you can search all the results of the entire Human Genome Project, check out the latest gene maps, and look up anything about any disease that has a genetic basis.

Not only that, but bioinformatics gives you ready access to powerful analytical tools — the kind the pros use. Gene hunters use these tools to compare human DNA sequences with those in other animals (see Chapter 8 for a rundown of critters whose DNA has been sequenced). As one of the next big things in genetics, bioinformatics provides the tools to catalog, keep track of, and analyze all the data generated by geneticists the world over. This data is then used for all the applications I cover in this book — from genetic counseling to cloning and beyond.

Gene Chips

Technology is at the heart of modern genetics, and one of the most useful developments in genetic technology is the *gene chip*. Also known as *microarrays,* gene chips allow researchers to quickly determine which genes are at work (that is, being expressed) in a given cell (see Chapter 11 for a full rundown on how your genes do their jobs).

Gene expression depends on messenger RNA (mRNA), which is produced through transcription (see Chapter 9). The mRNAs get tidied up and sent out into the cell cytoplasm to be translated into proteins (see Chapter 10 for how translation works to make proteins). The various mRNAs in each cell tell how many and exactly which of the thousands of genes are at work at any given moment. In addition, the number of copies of each mRNA conveys an index of the strength of gene expression (see Chapter 11 for more on gene expression). The more copies of a particular mRNA, the stronger the action of the gene that produced it.

Gene chips are grids composed of bits of DNA that are complementary to the mRNAs the geneticist expects to find in a cell (I explain the method used to detect the mRNAs in the first place in Chapter 16). It works like this: The bits of DNA are attached to a glass slide. All the mRNAs from a cell are passed over the gene chip, and the mRNAs bind to their DNA complements on the slide. Geneticists measure how many copies of a given mRNA attach themselves to any given spot on the slide to determine which genes are active and what their strength is.

Gene chips are relatively inexpensive to make and can each test hundreds of different mRNAs, making them a valuable tool for gene discovery and mapping. Scientists are also using microarrays to screen thousands of genes rapidly to pick up on mutations that cause diseases, as well as chromosome abnormalities (like those I describe in Chapter 15). One way they perform this screening is by comparing mRNAs from normal cells to those from diseased cells (such as cancer). By comparing the genes that are turned on or off in the two cell types, geneticists can determine what's gone wrong and how the disease may be treated.

Evolution of Antibiotic Resistance

Unfortunately, not all "next big things" are good. Antibiotics are used to fight diseases caused by bacteria. When penicillin (a common antibiotic) was developed, it was a wonder drug that saved thousands and thousands of lives. However, many antibiotics are nearly useless now because of the evolution of *antibiotic resistance*.

Bacteria don't have sex, but they still pass their genes around. They achieve this feat by passing around little circular bits of DNA called *plasmids*. Almost any species of bacteria can pass its plasmids on to any other species. Thus, when bacteria that are resistant to a particular antibiotic run into bacteria that aren't resistant, the exchange of plasmids endows the formerly susceptible bacteria with antibiotic resistance. Antibacterial soaps and the overprescribing of antibiotics make the situation worse by killing off all the nonresistant bacteria, leaving only the resistant kind behind.

Antibiotic-resistant bacteria are showing up not only in hospitals but also in natural environments. Farmers pump their animals full of antibiotics in an effort to keep them free from disease. Thus, antibiotic-resistant bacteria abound in farm sewage, and eventually, the runoff ends up in lakes, streams, and rivers that provide drinking water for humans. Many of those bacteria cause human diseases, and because they start off as antibiotic-resistant bacteria, treating illnesses that they cause is difficult. Meanwhile, scientists work to develop new, more powerful antibiotics in an effort to stay one step ahead of the bacteria.

Genetics of Infectious Disease

I'm guessing that you're too young to remember the flu epidemic of 1918 (I certainly am!). My aunt, who was a schoolteacher in 1918, told me that half the students at her tiny, rural Louisiana school died, along with the school's other teacher. All told, 20 million people worldwide died of the flu in that horrific epidemic. The virus was so deadly that people caught it in the morning and died the same day!

A frightening descendent of the virus that caused the 1918 pandemic is still around. Swine flu turned into a global pandemic in June 2009, affecting thousands of people around the world. Fortunately, this new virus, known as *H1N1,* is not as severe as its predecessor, causing only acute illness in most cases.

Influenza viruses frequently start out as bird diseases (usually carried in the guts of domestic poultry) that move from birds to a new host. Flu viruses pull off this transformation by picking up new genes from the DNA of their hosts or from other viruses. This means that the flu viruses are constantly evolving, changing their surface proteins to allow invasion into new hosts (like pigs and humans) and new organ systems (like airways and lungs).

The 2009 swine flu possesses genes from two different pig flu viruses (that is, viruses that cause influenza in the pigs themselves). Pigs are unusual in that they can contract flu from humans, birds, and one another. After they're infected, pig cells are capable of simultaneously hosting multiple viruses, which allows the viruses to acquire new genes very easily. It's not clear how this swine flu made the jump to humans, though, because pig-to-person transmission is very rare.

Bioterrorism

After September 11, 2001, terrorism moved to the forefront of many people's mind. Hot on the heels of the disaster in New York City was another threat in the form of anthrax-laced letters. (Anthrax is a deadly disease caused by a soil bacterium.) Opening junk mail in the United States went from merely annoying to potentially threatening.

Anthrax and other infectious organisms are potential weapons that can be used by terrorists — a form of warfare called *bioterror.* Suddenly, the researchers working on anthrax genetics — people who had toiled away in underfunded obscurity — were national treasures. U.S. government spending on efforts to counter the bioterror threat shot up. Since 2001, the United

States has spent roughly $50 billion on biodefense, including studies of infectious disease and measures aimed at protecting public health. As a result of these expenditures, scientists are also able to quickly identify the pathogens behind disease outbreaks unrelated to bioterror. For example, researchers identified a new species of Ebola (which causes a nearly always fatal form of hemorrhagic fever) during a 2008 outbreak in Uganda.

Critics have argued that the push for anti-bioterror research means that many important and more immediate problems go unsolved. Furthermore, the bad guys may not even have the technology needed to make the sophisticated biological weapons that big money is spent to counter. Meanwhile, new regulations make research harder to conduct. Scientists can no longer easily exchange biological samples, meaning that the experts can't always get the research materials they need to do their work.

DNA Bar Coding

You're probably familiar with the black and white codes on the packaging of everything you purchase, from peanuts to computers. The computer bar code allows stores to track inventory and pricing of every item they carry. One of the hottest topics in genetics is how the genetic code may be used in a similar way to identify and track living things.

The idea is pretty simple: Using genes in mitochondrial DNA (which you can read about in Chapter 6), scientists look for sequences unique to particular species. After they determine that a sequence reliably identifies a given species of animal, the "bar code" is registered in a database. So far, nearly 65,000 species have been matched with a DNA bar code.

Though the idea behind DNA bar coding is simple, the genetics behind it are amazingly complex. Because practically all organisms carry identical DNA sequences (you and a banana, for example, share over 90 percent of your DNA in common), finding sequences that match up with one, and only one, species has been difficult. For this reason, many researchers criticize the idea. In addition, closely related species rarely interbreed, but they do so often enough that genetic lines are too blurry to make bar coding them reliable. Nonetheless, DNA bar coding has tremendous potential. For example, some genetic sleuthing in 2009 showed that almost half of the New York City sushi restaurants sampled had mislabeled their fish and were even serving up endangered species such as bluefin tuna. Such information may eventually be used to protect fisheries from being overexploited and to protect consumers from being mislead.

Chapter 24

Ten Hard-to-Believe Genetics Stories

In This Chapter

▶ Animals that diverge from genetic norms
▶ Other amazing tales from the world of genetics

Think you've heard it all? Well, put on your seat belt and get ready for some wild, wacky, and woolly — but true — stories from the genetics lab.

Scrambled Genes: Platypuses Break All the Rules

It has a bill like a duck and lays eggs, but it has fur and produces milk. This creature, which hails from eastern Australia, also produces venom (like a snake) that's excreted by males from spurs on their hind limbs. Did I mention that this thing can swim and can sense electrical fields in the water to find fish? Is it a mammal? A bird?

It's a *platypus,* and not only does it boast a truly strange combination of bird, reptile, and mammal characteristics, but it also has one of the most bizarre systems for determining sex. Platypuses (or is it platypi?) have a whopping 10 sex chromosomes. Platypuses are diploid. Males have 21 pairs of chromosomes, plus 10 sex chromosomes: 5 Xs and 5 Ys. Females have 21 pairs of chromosomes (identical to those of males), plus 10 Xs. The fun doesn't stop there. The *SRY* gene that normally determines maleness in mammals — and yes, the platypus is considered a mammal — is absent. Instead, platypuses have a version of the bird sex-determining gene that's located on one of the 5 X chromosomes.

The sequence of the platypus genome reveals that this incredible creature has genes in common with reptiles, mammals, and birds. This finding suggests that the platypus (along with marsupials like kangaroos) is descended from a distant (like roughly 166 million years ago) reptilian ancestor of both mammals and birds. As scientists decipher the platypus's many genetic secrets, conservationists are working to preserve its habitat, which is in danger from climate change, among other threats.

What's in a Name?

Which is easier to remember: Lunatic Fringe or LFNG O-fucosylpeptide 3-beta-N-acetylglucosaminyltransferase? I'm guessing you picked the first one, right? When scientists discover a new gene, one of the perks is getting to name it. For many fruit fly genes, the names are witty, easy to remember, and informative. Take Groucho Marx, which calls to mind the bushy eyebrows and mustache of the comedian — the fly with the Groucho gene has lots of facial bristles. Cheap Date? A mutation that causes an unusual sensitivity to alcohol. Out Cold? When it's chilly, the fly with this mutation faints.

Not all scientists are laughing, however. The Human Genome Organization Gene Nomenclature Committee (whose own name could use a makeover) deemed some of these gene names "inappropriate, demeaning, and pejorative." That's because there are human versions of the same genes. To avoid hurt feelings among doctors and patients, the committee is renaming some of the genes from simple, easy to remember, and fun monikers to long, multi-syllabic, and, well, boring appellations. Goodbye Cheap Date; hello pituitary adenylyl cyclase-activating polypeptide gene.

Second Life

Long the stuff of science fiction, artificially created life forms may not be so far-fetched after all. In 2008, a team from the J. Craig Venter Institute announced that it had successfully synthesized an entire genome. (Venter is one of the pioneers of the Human Genome Project.) The newly created genome is a mock-up of the *Mycoplasma genitalium* gene sequence. To prevent the artificial version from leaking out and causing trouble, the researchers added a gene to make the bacteria dependent on an antibiotic for its growth and survival. They also added a John Hancock: The team members' and institute's names are spelled out in DNA code. Even with this advance, it's not artificial life. To see if the synthesized genome works as well as a real one, researchers will insert it into a cell. This step will improve scientists' understanding of what it takes to build genomes and how genes (real and synthesized) work.

Lousy Chromosomes

The human body louse, a tiny, bloodsucking bug, has a big claim to genetic fame. It boasts a total of 18 mitochondrial chromosomes. Most animals have only a single, circular mitochondrial chromosome that's transmitted from mother to offspring. But this louse, *Pediculus humanus,* has many mini-chromosomes that contain between one and three genes each. Pieced together, the 18 chromosomes (more or less) add up to the usual one chromosome and its complement of 37 genes.

Not Yourself: DNA Chimeras

Imagine being told that, as a mom, the children you thought were yours, the ones you gave birth to, aren't your own. This is exactly what happened to a patient at Beth Israel Deaconess Medical Center in Boston, Massachusetts. While searching for a possible donor for a kidney transplant, the woman — we'll call her Sally — was told that her two sons weren't hers, although they were fathered by her husband.

After a lot of speculation and DNA testing, researchers discovered that Sally's tissues have two genetic signatures that are so distinct from each other that they could be from two different people. And that's exactly where they came from: two embryos, nonidentical twin girls, that fused during development to make one person — Sally. That makes Sally a *tetragametic chimera,* or, to put it more simply, a fusion of two eggs fertilized by two sperm (that's the *tetra-gametic* or *four-gametes* part). The word *chimera* comes from the name of a mythical creature that had a lion's head and a goat's body.

It turns out that chimeras may be common, from a genetic standpoint. Studies show that many women carry their mother's white blood cells, which apparently cross from mom to child during pregnancy. These cells act to keep the growing fetus healthy and then continue to divide and fight disease through-out the individual's lifetime, all the while containing an entirely different person's DNA fingerprint. As more evidence comes in, this form of chimerism may turn out to be the norm instead of the exception, meaning that all of us may not be quite ourselves after all.

Genes Even a Mother Could Love

Motherly love does more than make you feel good. It also affects your genes. Back in 2004, a group of geneticists discovered that in rats, individuals that as pups were well cared for by their moms had epigenetic patterns that differed from rats that were neglected when young. These changes altered the DNA in the rats' brain tissues, making them more or less susceptible to stress depending on mom's caring behavior.

The researchers then wondered if humans have similar epigenetic changes in their brains. They compared the brains of people who were abused as children and later committed suicide to the brains of people without histories of abuse who died of other causes. The findings were stunning. The suicide victims had genetic changes to their brain tissue's DNA, and the changes were confined only to the tissues of the brain where stress hormones are regulated. This means that the expression of your genes may be far more malleable and sensitive to your experiences than previously realized. It also opens the possibility of uncovering causes of mental illness, as well as treatments and prevention.

One Gene to Rule Them All

A single gene controls pain perception. It was discovered when a boy in Pakistan came to the attention of medical authorities. As a street performer, the child walked over hot coals and stabbed himself with knives — not with trickery, but for real. It turned out that the child, and many of his relatives, felt no pain of any kind. Researchers tracked down a single gene mutation that controls all perception of pain. This is exciting news because it means that scientists may be able to formulate a drug to target the gene, controlling pain for people who suffer from debilitating conditions.

Why Alligators May Outlast Us All

Alligators are survivors. They live in nasty places, eat dead stuff, and don't seem to be bothered by much of anything. In most creatures, an individual must be exposed to a disease-causing agent to become immune to it. Not so for alligators. Recent research shows that in addition to their toothy smiles, these massive reptiles have blood proteins that fight off infections from bacteria, fungi, and viruses, even without previous exposure to the offending microorganism. In tests, alligator blood proteins kill even antibiotic-resistant "super bugs" that cause thousands of human deaths each year. Scientists are now working to determine how to develop drugs based on the chemical structure of gator blood.

Do-It-Yourself Genetics

Genetic testing used to be reserved for special cases and dire circumstances. Not anymore. Now, getting yourself tested for all manner of genetic information —

from the possibility of cancers to a predisposition for male-pattern baldness — is a simple matter. Some enterprising people have even turned to doing genetic testing right in the comfort of their own home. Using castoff lab equipment purchased from that amazing online marketplace, eBay, and lab protocols downloaded from the Internet, it's relatively simple to probe your own genes.

 There's even an organization called DIYbio (www.diybio.org) whose members have gatherings like DNA extraction parties (I swear I'm not making this up) and provide info about and resources for taking biotechnology home . . . and using it.

Making Something Useful Out of Junk

 Researchers are finding that "junk DNA," proving to be far more functional and essential than ever before, is crucial for turning genes on and off. In addition, it may have been junk DNA that flipped the switch and gave primates like us the ability to grasp tools and walk upright. A particular section of noncoding DNA called HACNS1 kicks off genes that appear to control development of thumbs and big toes. These findings have led scientists to believe that junk DNA may be just as important in evolution as the genes themselves.

Glossary

adenine: Purine base found in DNA and RNA.

allele: Alternative form of a gene.

amino acid: Unit composed of an amino group, a carboxyl group, and a radical group; amino acids link together in chains to form polypeptides.

anaphase: Stage of cell division in mitosis when replicated chromosomes (as chromatids) separate. In meiosis, homologous chromosomes separate during anaphase I, and replicated chromosomes (as chromatids) separate during anaphase II.

aneuploidy: Increase or decrease in the number of chromosomes; a deviation from an exact multiple of the haploid number of chromosomes.

anticipation: Increasing severity or decreasing age of onset of a genetic trait or disorder with successive generations.

anticodon: The three nucleotides in a tRNA (transfer RNA) complementary to a corresponding codon of mRNA.

antiparallel: Parallel but running in opposite directions; orientation of two complementary strands of DNA.

apoptosis: Normal process of regulated cell death.

autosome: A nonsex chromosome.

backcross: Cross between an individual with an F1 genotype and an individual with one of the parental (P) genotypes.

bacteriophage: Virus that infects bacterial cells.

base: One of the three components of a nucleotide. DNA and RNA have four bases.

cell cycle: Repeated process of cell growth, DNA replication, mitosis, and cytokinesis.

centromere: Region at the center of a chromosome that appears pinched during metaphase; where spindle fibers attach during mitosis and meiosis.

chromatid: One half of a replicated chromosome.

chromosome: Linear or circular strand of DNA that contains genes.

codominance: When heterozygotes express both alleles equally.

codon: Combination of three nucleotides in an mRNA that correspond to an amino acid.

complementary: Specific matching of base pairs in DNA or RNA.

consanguineous: Mating by related individuals.

crossing-over: Equal exchange of DNA between homologous chromosomes during meiosis.

cytokinesis: Cell division.

cytosine: A pyrimidine base found in DNA and RNA.

ddNTP: Dideoxyribonucleotide; identical to dNTP but lacking an oxygen at the 3' site. Used in DNA sequencing.

deamination: When a base loses an amino group.

degenerate: Property of the genetic code whereby some amino acids are encoded by more than one codon.

deletion: Mutation resulting in the loss of one or more nucleotides from a DNA sequence.

denaturation: Melting bonds between DNA strands, thereby separating the double helix into single strands.

depurination: When a nucleotide loses a purine base.

dihybrid cross: Cross between two individuals who differ at two traits or loci.

diploid: Possessing two copies of each chromosome.

DNA: Deoxyribonucleic acid; the molecule that carries genetic information.

dNTP: Deoxyribonucleotide; the basic building block of DNA used during DNA replication consisting of a deoxyribose sugar, three phosphate molecules, and one of four nitrogenous bases.

dominant: An allele or phenotype that completely masks another allele or phenotype. The phenotype exhibited by both homozygotes and heterozygotes carrying a dominant allele.

epigenetics: Changes in gene expression and phenotype caused by characteristics of DNA outside the genetic code itself.

epistasis: Gene interaction in which one gene hides the action of another.

eukaryote: Organism with a complex cell structure and a cell nucleus.

euploid: Organism possessing an exact multiple of the haploid number of chromosomes.

exon: Coding part of a gene.

expressivity: Variation in the strength of traits.

F1 generation: First generation offspring of a specific cross.

F2 generation: Offspring of the F1 generation.

gamete: Reproductive cell; sperm or egg cell.

gene: Fundamental unit of heredity. A specific section of DNA within a chromosome.

genome: A particular organism's full set of chromosomes.

genotype: The genetic makeup of an individual. The allele(s) possessed at a given locus.

guanine: Purine base found in DNA and RNA.

gyrase: Enzyme that acts during DNA replication to prevent tangles from forming in the DNA strand.

haploid: Possessing one copy of each chromosome.

helicase: Enzyme that acts during DNA replication to open the double helix.

heterozygote: Individual with two different alleles of a given gene or locus.

homologous chromosomes: Two chromosomes that are identical in shape and structure and carry the same genes. Diploid organisms inherit one homologous chromosome from each parent.

homozygote: Individual with two identical alleles of a given gene or locus.

insertion: Mutation resulting in the addition of one or more nucleotides to a DNA sequence.

interphase: Period of cell growth between divisions.

intron: Noncoding part of a gene. Intervening sequences that interrupt exons.

ligase: Enzyme that acts during replication to seal gaps created by lagging strand DNA synthesis.

linkage: Inheriting genes located close together on chromosomes as a unit.

locus: A specific location on a chromosome.

meiosis: Cell division in sexually reproducing organisms that reduces amount of genetic information by half.

metaphase: Stage of cell division when chromosomes align along the equator of the dividing cell.

mitosis: Simple cell division without a reduction in chromosome number.

nucleotide: Building block of DNA; composed of a deoxyribose sugar, a phosphate, and one of four nitrogenous bases.

P generation: Parental generation in a genetic cross.

penetrance: Percentage of individuals with a particular genotype that express the trait.

phenotype: Physical characteristics of an individual.

polypeptide: Chain of amino acids that form a protein.

prokaryote: Organism with a simple cell structure and no cell nucleus.

prophase: Stage of cell division when chromosomes contract and become visible and nuclear membrane begins to break down. In meiosis, crossing-over takes place during prophase.

purine: Compound composed of two rings.

pyrimidine: Chemicals that have a single, six-sided ring structure.

recessive: A phenotype or allele exhibited only by homozygotes.

replication: Process of making an exact copy of a DNA molecule.

RNA: Ribonucleic acid; the single-stranded molecule that transfers information carried by DNA to the protein-manufacturing part of the cell.

telomere: Tip of a chromosome.

telophase: Stage of cell division when chromosomes relax and the nuclear membrane re-forms.

thymine: Pyrimidine base found in DNA but not RNA.

totipotent: A cell that can develop into any type of cell.

uracil: Pyrimidine base found in RNA but not DNA.

zygote: Fertilized egg resulting from the fusion of a sperm and egg cell.

Index

Business/Accounting & Bookkeeping

Bookkeeping For Dummies
978-0-7645-9848-7

eBay Business
All-in-One For Dummies,
2nd Edition
978-0-470-38536-4

Job Interviews
For Dummies,
3rd Edition
978-0-470-17748-8

Resumes For Dummies,
5th Edition
978-0-470-08037-5

Stock Investing
For Dummies,
3rd Edition
978-0-470-40114-9

Successful Time
Management
For Dummies
978-0-470-29034-7

Computer Hardware

BlackBerry For Dummies,
3rd Edition
978-0-470-45762-7

Computers For Seniors
For Dummies
978-0-470-24055-7

iPhone For Dummies,
2nd Edition
978-0-470-42342-4

Laptops For Dummies,
3rd Edition
978-0-470-27759-1

Macs For Dummies,
10th Edition
978-0-470-27817-8

Cooking & Entertaining

Cooking Basics
For Dummies,
3rd Edition
978-0-7645-7206-7

Wine For Dummies,
4th Edition
978-0-470-04579-4

Diet & Nutrition

Dieting For Dummies,
2nd Edition
978-0-7645-4149-0

Nutrition For Dummies,
4th Edition
978-0-471-79868-2

Weight Training
For Dummies,
3rd Edition
978-0-471-76845-6

Digital Photography

Digital Photography
For Dummies,
6th Edition
978-0-470-25074-7

Photoshop Elements 7
For Dummies
978-0-470-39700-8

Gardening

Gardening Basics
For Dummies
978-0-470-03749-2

Organic Gardening
For Dummies,
2nd Edition
978-0-470-43067-5

Green/Sustainable

Green Building
& Remodeling
For Dummies
978-0-470-17559-0

Green Cleaning
For Dummies
978-0-470-39106-8

Green IT For Dummies
978-0-470-38688-0

Health

Diabetes For Dummies,
3rd Edition
978-0-470-27086-8

Food Allergies
For Dummies
978-0-470-09584-3

Living Gluten-Free
For Dummies
978-0-471-77383-2

Hobbies/General

Chess For Dummies,
2nd Edition
978-0-7645-8404-6

Drawing For Dummies
978-0-7645-5476-6

Knitting For Dummies,
2nd Edition
978-0-470-28747-7

Organizing For Dummies
978-0-7645-5300-4

SuDoku For Dummies
978-0-470-01892-7

Home Improvement

Energy Efficient Homes
For Dummies
978-0-470-37602-7

Home Theater
For Dummies,
3rd Edition
978-0-470-41189-6

Living the Country Lifestyle
All-in-One For Dummies
978-0-470-43061-3

Solar Power Your Home
For Dummies
978-0-470-17569-9

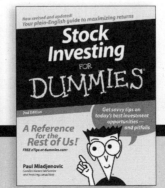